THE SIX SECRET TEACHINGS
ON THE WAY OF STRATEGY

The
Six Secret
Teachings
on the Way
of Strategy

TRANSLATED BY

RALPH D. SAWYER

With the Collaboration of

MEI-CHÜN LEE SAWYER

SHAMBHALA
Boston & London
1997

SHAMBHALA PUBLICATIONS
Horticultural Hall
300 Massachusetts Avenue
Boston, Massachusetts 02115
http://www.shambhala.com

© 1993, 1995, 1996 by Ralph D. Sawyer

This translation was previously published in.
The Seven Military Classics of Ancient China and is
reprinted by arrangement with Westview Press.

9 8 7 6 5 4 3 2 1

First Edition
Printed in the United States of America
♾ This edition is printed on acid-free paper that meets
the American National Standards Institute z39.48 Standard.
Distributed in the United States by Random House, Inc.,
and in Canada by Random House of Canada Ltd

LIBRARY OF CONGRESS CATALOGING-IN-PUBLICATION DATA
Lü, Shang.
[Liu t'ao. English]
The six secret teachings on the way of strategy: a manual
from ancient China / translated by Ralph D. Sawyer, with
the collaboration of Mei-chün Lee Sawyer
p.   cm.
ISBN (invalid) 1-57062-314-0 (alk. paper)
1. Military art and science—Early works to 1800. I. Sawyer,
Ralph D.   II. Sawyer, Mei-chün.
U101.L83L813 1997   96-32871
355.02—dc20   CIP

Contents

V   Leopard Secret Teaching

VI   Canine Secret Teaching

Preface

The Six Secret Teachings, a remarkable work of strategy and tactics dating from the late Warring States period, is the most comprehensive of the so-called *Seven Military Classics*, collected and made canonical in the Sung dynasty. Although *The Six Secret Teachings* has never enjoyed the degree of respect accorded Sun-tzu's *Art of War* by non-military thinkers in China, and was disparaged by effete bureaucrats who found it unthinkable that one of China's greatest figures was apparently a political realist as well as a military sage, real commanders and those interested in preserving the state amid the turmoil that repeatedly engulfed her over the centuries valued it greatly and extensively employed its teachings. The most extensive of the military writings, it explores numerous fundamental topics from an essentially humanistic, Taoistically oriented perspective while also providing a compendium of useful, frequently unorthodox tactics for concrete application. Interspersed among its pages may be found many of man's most pressing concerns, including social organization, development of virtue, creation of a productive society, and the role of the military. Throughout, the essential vision is one of maneuver warfare—battlefield encounters designed to minimize losses while still retaining the martial spirit and capability to prevail in history's perhaps most turbulent era, the Warring States.

For this edition of *The Six Secret Teachings* (previously translated in my *Seven Military Classics of Ancient China*) a new introduction has been prepared that, while including essential historical material necessary to fully understand and appreciate the text, delves into the Taoist prescriptive underlying much of the work and sets the teachings within China's humanitarian thrust. The Chinese military writers, far more than the bureaucrats that came to dominate the country's administration, realized that—in Sun-tzu's words—"warfare is the greatest affair of state, the basis of life or death, the Tao to survival or extinction." Moreover, they well understood, as the *Ssu-ma Fa* states, that "Those who love warfare will invariably perish, those who forget warfare will certainly be endangered." The difficulty of protecting oneself from evil during perverse moments must be balanced with concerns for nurturing and sustaining all humanity; therefore, both compete for limited resources in every complex society. Accordingly, the introduction emphasizes *The Six Secret Teachings'* fundamental views within the intellectual and martial ferment of the Warring States period rather than simply focusing upon military affairs and tactical concerns, as was done in my *Seven Military Classics of Ancient China.*

It hardly need be noted that *The Six Secret Teachings*, like all the military classics, has in recent years enjoyed an astounding revival not just in China but throughout the Far East, including Japan and Southeast Asia. Numerous popular versions have appeared, some sloppy interpretations oriented toward resolving age-old problems in business and human relations, others more complete and authoritative. All have sold well into six figures, suggesting the consternation that has accompanied Asia's rapid immersion in the throes of modern economic realities whereby traditional—sometimes repressive, sometimes beneficial—value systems have been discarded and populaces wrenched from their comforting, if protective, confines to confront a new material age and competitive environment. Thus *The Six Secret Teachings* remains relevant not only to the human quest for understanding in general, but also to the events unfolding in an ever-evolving contemporary Asia.

Appreciation is due to Shambhala Publications for their wisdom in making this work available to a wider readership, and especially to Peter Turner for suggestions that stimulated us to ponder its Taoist threads for the expanded introduction. Our thanks also to Westview Press for supporting the initial work and acceding to its publication in this single-volume format, and to Peter Kracht and Max Gartenberg for their extensive contributions and guidance over the years.

Ralph D. Sawyer

A Note on Pronunciation

Unfortunately, neither of the two commonly employed orthographies makes the pronunciation of romanized Chinese characters easy. Each system has its stumbling blocks and we remain unconvinced that the Pinyin *qi* is inherently more comprehensible to unpracticed readers than the Wade-Giles *ch'i*, although it is certainly no less comprehensible than *j* for *r* in Wade-Giles. However, as many of the important terms may already be familiar and previous translations of Sun-tzu's *Art of War* have mainly used Wade-Giles, we have opted to employ it throughout our works, including *The Six Secret Teachings*. Well-known cities, names, and books—such as "Peking"—are retained in their common form, and books and articles published with romanized names and titles also appear in their original form.

As a crude guide to pronunciation we offer the following notes on the significant exceptions to normally expected sounds:

> *t,* as in *Tao:* without apostrophe, pronounced like *d*
> *p,* as in *ping:* without apostrophe, pronounced like *b*
> *ch,* as in *chuang:* without apostrophe, pronounced like *j*
> *hs,* as in *hsi:* pronounced *sh*
> *j,* as in *jen:* pronounced like *r*

Thus, the name of the famous Chou dynasty is pronounced as if written "jou" and sounds just like the English name "Joe."

TRANSLATOR'S INTRODUCTION

Historical Context

The Six Secret Teachings purportedly records the strategic advice and tactical instructions given in the mid-eleventh century BCE by a famous general known as T'ai Kung (an honorific title) to Kings Wen and Wu, the Chou dynasty's illustrious founders. Although the present book unquestionably dates from the late Warring States period—as will be discussed at the end of the introduction—some traditionalists still believe it reflects the mature heritage of military studies found in the strong state of Ch'i, and therefore preserves at least vestiges of the oldest strata of Chinese military thought. The T'ai Kung, a pivotal historical figure to whom *The Six Secret Teachings* is nominally attributed, has been honored throughout Chinese history as the first famous general and progenitor of strategic studies. Furthermore, during the early years of the T'ang dynasty—whose founders were much imbued with a martial spirit and valued military affairs equally with civil ones—he was even accorded his own state temple as the martial patron, thereby at least temporarily attaining officially sanctioned status approaching that of Confucius, the revered civil patron.

A complete work that discusses not only strategy and tactics, but also proposes the government measures necessary for forging effective state control and attaining national prosperity, *The Six Secret Teachings* is deliberately grounded upon—or perhaps projected back into—monumental historical events. The Chou kings presumably implemented many of these policies, thereby enabling them to develop their agricultural and population bases, gradually expand their small border domain, civilize and transform the disparate peoples brought under its sway, and secure the allegiance of the populace until

I

they could finally launch the decisive military campaign that defeated the powerful Shang dynasty and overturned its six hundred years of domination.

Accordingly, *The Six Secret Teachings* is the only ancient Chinese military classic written from the perspective of revolutionary activity, for the Chou's aim was nothing less than a dynastic revolution. Attaining their objective required perfecting themselves in the era's measures and technologies, as well as systematically developing policies, strategies, and even battlefield tactics not previously witnessed in Chinese history. From their position on the realm's periphery the Chou kings were compelled to ponder employing limited resources and restricted forces to attack a vastly superior, well-entrenched foe whose campaign armies alone probably outnumbered their entire population. In contrast, the other strategic writings focus upon managing military confrontations between states of somewhat comparable strength, with both sides starting from relatively common military and government infrastructures. Furthermore, while nearly all the military texts vociferously advocate the fundamental policies of "enriching the state [through agriculture] and strengthening the army," many tend to emphasize strategic analyses and battlefield tactics rather than the fundamental measures necessary to even create the possibility of confrontation.

The final, epoch-making clash between the Chou and Shang, as envisioned by the Chou themselves and idealistically portrayed in later historical writings, set the moral tone and established the parameters for China's concept of dynastic cycles. The archetypal battle of Virtue and evil—the benevolent and righteous Chou monarchy acting on behalf of all people everywhere against the Shang tyrant and his coterie of parasitic supporters—truly commenced with their conflict. The Shang's earlier conquest of the Hsia, while portrayed as similarly conceived—King T'ang, the founder of the Shang dynasty, having cultivated his Virtue, pursued benevolent policies, and garnered his strength on the fringe of the Hsia empire until finally engaging in a decisive battle—occurred before the advent of written language, and thus remained only a legend even in antiquity. However, the Chou's determined effort to free the realm from the yoke of suffering and establish a rule of Virtue and benevolence became the inspirational

essence for China's subsequent moral self-perception. As dyna
decayed and their rulers became morally corrupt and increas......,
ineffectual, new champions of righteousness would appear to force-
fully confront the government's oppressive forces, rescue the populace
from imminent doom, and return the state to benevolent policies.
While temporary resurgences might be effected by vigorous kings, and
were even achieved during the Shang, the process of decline inex-
orably led to the dynasty's eventual vanquishment, generally accompanied by widespread misery and suffering.

As portrayed in such historical writings as the *Shih Chi*, in accord
with this moral imperative and in response to the plight of the people,
the Shang had initially ascended to power by vanquishing the last evil
ruler of the reigning Hsia dynasty. After generations of often glorious,
although theocratic, rule the successive Shang emperors—perhaps
due to their splendid isolation and constant indulgence in myriad
pleasures—are portrayed as having similarly become ever less virtuous
and capable. Their moral decline continued largely unabated until
King Chou, the final ruler whom history has depicted as evil incar-
nate. The many perversities attributed to him include imposing heavy
taxes; forcing the people to perform onerous labor services, mainly to
provide him with lavish palaces and pleasure centers; interfering with
the people's agricultural practices, thereby causing widespread hunger
and deprivation; extreme debauchery, including drunkenness, orgies,
and violence; the brutal murder of innumerable people, particularly
famous men of Virtue and loyal court officials; and the development
and widespread infliction of inhuman punishments. Unfortunately, as
the following brief excerpt from the "Shang Annals" in the *Shih Chi*
notes, the king was also talented, powerful, and fearsome:

> In discrimination and natural ability Emperor Chou was quick and
> acute; his sight and hearing were extremely sensitive; and his
> strength and physical skills surpassed other men. He could slay a
> fierce animal with his bare hands; his knowledge easily warded off
> criticism; and his verbal skills readily adorned his errors. He
> boasted about his own ability to his ministers; he was haughty to
> all the realm because of his reputation; and he believed that
> everyone was beneath him. He loved wine, debauched himself
> in music, and was enamored of his consorts. He especially loved

Ta Chi, and invariably followed her words. Thus he had Shih Chuan create new licentious sounds, the Pei-li dance of licentious women, and the lewd music of "fluttering down." He made the taxes heavier in order to fill the Deer Tower with coins, and stuffed the Chu-ch'iao storehouses with grain. He increased his collections of dogs, horses, and unusual objects until they overflowed the palace buildings. He expanded the Sha-ch'iu garden tower, and had a multitude of wild animals and flying birds brought there. He was disrespectful to ghosts and spirits. He assembled numerous musicians and actors at the Sha-ch'iu garden; made a lake of wine and a forest of hanging meat, amidst which naked men and women pursued each other; and conducted drinking feasts throughout the night. The nobility known as the hundred surnames looked toward him with hatred, and some of the feudal lords revolted.

According to traditional historical sources, the Chou state was dramatically established when Tan Fu, their leader, personally emigrated over the mountains south into the Wei River valley to avoid endangering his people and subsequently abandoned so-called barbarian customs to embrace the agricultural destiny of his ancestors. These actions immediately characterized him as a paragon of Virtue, and endowed the Chou, and subsequently China, with a sedentary, agrarian character. As recorded in the *Shih Chi*:

> The Ancient Duke, Tan Fu, again cultivated the agricultural occupation of Hou Chi and Duke Liu, accumulated his Virtue and practiced righteousness, and thus the people of the state all supported him. The Hsün-yü of the Jung and Ti steppe peoples attacked them, wanting to seize their wealth and material goods, so Tan Fu yielded them. After that the Hsün-yü again attacked, wanting to take the land and people. The people were all enraged and wanted to fight. The Ancient Duke said, "Establishing a ruler should be to the people's advantage. Now these barbarians are attacking and waging war with us because they want my land and people. What difference is there if the people are with them, or with me? The people want to fight because of me, but I cannot bear to slay people's fathers and sons in order to rule them." Then, with his relatives, he went to Pin, forded the Ch'i and Chu Rivers,

crossed over Mount Liang, and stopped below Mount Ch'i. Again the people of Pin, supporting their aged and carrying their weak, all flocked to join the Ancient Duke below Mount Ch'i. When nearby states heard of the Ancient Duke's benevolence, many also gave their allegiance. Thereupon the Ancient Duke abandoned the Chou's former barbarian customs, had walls and buildings constructed, and established cities to have the various peoples dwell separately. He also set up officials for the five offices. The people all sang songs and took pleasure in these works, praising his Virtue.

It has been suggested that the Chou easily developed alliances with nearby peoples, including disenchanted Hsia groups now subjugated by the Shang, because of their agricultural heritage. In reflecting the spirit of their remote ancestor Hou Chi (Lord of Millet) and perpetuating the Hsia's agricultural offices, the Chou had dispatched advisors to instruct other peoples and states in basic farming practices and seasonally appropriate activities for many years. Their efforts not only garnered them goodwill and respect, but also provided an opportunity to acquire a thorough knowledge of the inhabitants, customs, and terrain outside the Wei River valley.

However, Chi Li, Tan Fu's third son—who ascended the throne through the virtuous deference of his two elder brothers—aggressively waged successful campaigns against neighboring peoples and rapidly expanded their base of power. The Shang initially recognized his achievements and sanctioned his actions, granting him the title of earl, but eventually imprisoned him and he subsequently died at Shang hands despite having married into the royal house. While the history of Shang-Chou relations remains somewhat unclear, awaiting further archaeological discoveries, several other members of the Chou royal house, including King Wen, seem to have married Shang princesses. Moreover, generations before the Chou had migrated into the Wei River valley, commencing about the time of the famous Wu Ting, successive Shang kings had conducted military expeditions to subjugate the Chou. Furthermore, they had frequently hunted in the Chou's domain, but apparently grew apprehensive and abandoned the practice as the latter's might visibly increased.

In his later years King Wen of the Chou had also been imprisoned by Emperor Chou, the last tyrannical Shang ruler, for presuming to offer loyal remonstrance, possibly for several years during which he reputedly devoted himself to serious contemplation, ordered the sixty-four hexagrams of the *I Ching*, and appended the judgments to the hexagrams—all activities befitting a future cultural legend. Eventually he was freed, but not before his family and other virtuous men from across the realm offered overwhelmingly lavish bribes. In fact, the gifts thus assembled were apparently so impressive that King Wen, who throughout had continued to profess his submission and fealty to the Shang, was designated as the Western Duke or Lord of the West. When the title was conferred he was presented with a bow, arrows, and axes symbolic of the attendant military responsibilities that ironically required him to actively protect the imperial domain from external challenges. He immediately returned to his small state on the western fringe of the Shang empire, where the remoteness of the Wei River valley proved immensely advantageous. Dwelling in essentially barbarian territory, the Chou suffered the constant stimulus of vigorous military activity, but also enjoyed the harvests of a fertile area and the secrecy that relative isolation provided. Because King Wen could therefore implement many effective policies to foster the state's material and social strength without attracting undue attention, the Chou had the luxury of another seventeen years to prepare for the ultimate confrontation with the Shang.

THE T'AI KUNG

Into this state of Chou—insignificant when compared with the strength and expanse of the mighty Shang, which continued to exercise at least nominal control over some three thousand small states and fiefs—came the eccentric T'ai Kung, whose personal name was Chiang Shang. An elderly, mysterious figure whose early life was virtually unknown even then, he had perhaps found the Shang ruler insufferable and feigned madness to escape court life and its stultifying power. He disappeared only to resurface in the Chou countryside at the apocryphal age of seventy-two and become instrumental in Chou affairs. After faithfully serving the Chou court for approximately twenty years subsequent to his first encounter with King Wen, he was enfeoffed as King of Ch'i following the great conquest, as much to stabilize the eastern area and perhaps remove him as a military threat as to reward him for his efforts.

In all the stories about the T'ai Kung found in Warring States and later writings, he is invariably portrayed as old, insignificant, and impoverished. For example, the *Shuo Yüan,* a compilation of miscellaneous material, frequently employs his late, meteoric rise to power after an undistinguished career to illustrate the principle that talent and merit alone are inadequate for success unless one encounters the proper moment, the right opportunity, and wise mentors. One passage states: "When the T'ai Kung was fifty he sold food in Chi-chin; when he was seventy he butchered cows in Chao-kao; so if when he was ninety he commanded the army for the Son of Heaven, it was because he met King Wen." Another passage concludes: "His lands were inadequate to repay the cost of the seeds, his catch from fishing insufficient to cover the cost of the nets, but for governing All under Heaven he had more than enough wisdom." Finally, he was also identified as a boatman and common laborer, and even as someone whose wife had thrown him out!

Apart from his storied longevity and variegated past, his initial interview with King Wen is marked by the mythic aura that frequently seems to characterize predestined meetings between great historical figures. As recorded in *The Six Secret Teachings,* the Grand Historian had noted signs portending the appearance of a great Worthy and accordingly informed King Wen. The king therefore observed a vegetarian fast for three days to morally prepare for the meeting and ensure attaining the proper spiritual state of mind. When he finally encountered him the T'ai Kung quickly broached the ultimate subject of revolution, of overthrowing the Shang, by responding in allegorical terms to the king's inquiry about fishing. Immediately thereafter he abandoned metaphors to openly advise the king that the realm, the entire world, could be taken with the proper humanitarian measures and an effective government. Surprised by his directness, although probably assuming it was the working of Heaven, the king immediately acknowledged the T'ai Kung as the true Sage critical to realizing their dreams, and resolved to overthrow the Shang dynasty. Thereafter the T'ai Kung served as advisor, teacher, confidant, Sage, military strategist, and possibly commander-in-chief of the armed forces to Kings

Wen and Wu over the many years necessary before final victory could be realized.

The *Shih Chi* chapter reprising the history of the state of Ch'i contains a brief biography of its founder, the T'ai Kung, that provides additional information as well as recording the developments leading to the famous interview purportedly preserved in the initial chapter of *The Six Secret Teachings* itself:

"T'ai Kung Wang, properly named Lü Shang, was a native of the Eastern Sea area. One of his ancestors had served as a labor director and meritoriously assisted the legendary Yü in pacifying the flooding waters. In the interval between Emperor Shun and the Hsia dynasty he was therefore enfeoffed at Lü, or perhaps at Shen, and surnamed Chiang. During the Hsia and Shang dynasties some of the sons and grandsons of his collateral lines were enfeoffed at Lü and Shen, some were commoners, and Lü Shang—the T'ai Kung—was their descendant. His original surname was Chiang, but he was subsequently surnamed from his fief, so was called Lü Shang.

"Impoverished and in straits, Lü Shang was already old when, through fishing, he sought out King Wen, Lord of the West. The Lord of the West was about to go hunting, and so divined about the prospects for success. What the diviner said was: 'What you will obtain will be neither dragon nor serpent, neither tiger nor bear. What you will obtain is an assistant for a hegemon or king.'

"Thereupon the Lord of the West went hunting, and indeed encountered the T'ai Kung on the sunny side of the Wei River. After speaking with him he was greatly pleased and said: 'My former lord, the T'ai Kung, said, "There should be a Sage who will come to Chou, and Chou will thereby flourish." Are you truly this Sage or not? My T'ai Kung looked out [*wang*] for you for a long time.' Thus he called him T'ai Kung Wang, and returned together with him in his carriage, establishing him as strategist.

"Someone has said: 'The T'ai Kung had extensive learning, and once served King Chou of the Shang. King Chou lacked the Tao, so he left him. He traveled about exercising his persuasion on the various feudal lords but did not encounter anyone suitable, and in the end returned west with the Lord of the West.'

"Someone else has said: 'Lü Shang was a retired scholar who

had hidden himself on the seacoast. When the Lord of the West was confined at Yu-li, San-i Sheng and Hung Yao, having long known him, summoned Lü Shang. Lü Shang also said, "I have heard that the Lord of the West is a Worthy, and moreover excels at nurturing the old, so I guess I will go there." The three men sought out beautiful women and rare objects on behalf of the Lord of the West, and presented them to the Shang king in order to ransom the Lord of the West. The Lord of the West was thereby able to go out and return to his state.'

"Although the ways they say Lü Shang came to serve the Lord of the West differ, still the essential point is that he became strategist to Kings Wen and Wu. After the Lord of the West was extricated from Yu-li and returned to Chou, he secretly planned with Lü Shang and cultivated his Virtue in order to overturn Shang's government. The T'ai Kung's affairs were mostly concerned with military authority and unorthodox stratagems, so when later generations speak about armies and the Chou's secret tactical advantage of power, they all honor the T'ai Kung for making the fundamental plans.

"The Lord of the West's government was equitable, even extending to settling the conflict between the Yü and Jui peoples. Thus the *Book of Odes* refers to the Lord of the West as King Wen once he received the Mandate of Heaven. He successfully attacked the states of Ch'ung, Mi-hsü, and Chüan-i, and constructed a great city at Feng. If All under Heaven were to be divided into thirds, two thirds had already given their allegiance to the Chou. The T'ai Kung's plans and schemes occupied the major part.

"When King Wen died, King Wu ascended the throne. In his ninth year, wanting to continue King Wen's task, he mounted a campaign in the east to observe whether the feudal lords would assemble or not. When the army set out, the T'ai Kung wielded the yellow battle-ax in his left hand and grasped the white pennon in his right, in order to swear the oath:

> Ts'ang-ssu ! Ts'ang-ssu !
> Unite your masses of common people
> With your boats and oars.
> Those who arrive after will be beheaded.

Thereafter he went to Meng-chin. The number of feudal lords who assembled of their own accord was eight hundred. The feudal lords all said, 'King Chou can be attacked.' King Wu replied, 'Not yet.' He had the army return and made the Great Oath with the T'ai Kung.

"After they had remained in Chou for two more years, King Chou of the Shang killed Prince Pi-kan and imprisoned the Worthy Chi-tzu. King Wu, wanting to attack the Shang, performed divination with the tortoise shell to observe the signs. They were not auspicious, and violent wind and rain arose. The assembled dukes were all afraid, but the T'ai Kung stiffened them to support King Wu. King Wu then went forth.

"In the eleventh year, the first month, on the day *chia-tzu,* the king swore an oath at Mu-yeh and attacked King Chou of the Shang. The Shang army was completely defeated, so King Chou turned and ran off, and then ascended the Deer Tower. King Wu's forces pursued and beheaded King Chou. On the morrow King Wu was established at the altars: the dukes presented clear water; K'ang Shu-feng of Wei spread out a variegated mat; the T'ai Kung led the sacrificial animals; and the Scribe Yi chanted the prayers, in order to announce to the spirits the punishment of King Chou's offenses. They distributed the money from the Deer Tower, and gave out the grain in the Chü-ch'iao granary, in order to relieve the impoverished people. They enfeoffed Pi-kan's grave and released Chi-tzu from imprisonment. They removed the nine cauldrons of authority, rectified the government of Chou, and began anew with All under Heaven. The T'ai Kung's plans occupied the major part.

"Thereupon King Wu, having already pacified the Shang and become King of All under Heaven, enfeoffed the T'ai Kung at Ying-ch'iu in Ch'i. The T'ai Kung went east to his state, staying overnight on the road and traveling slowly. The innkeeper said, 'I have heard it said that time is hard to get but easy to lose. Our guest sleeps extremely peacefully. Probably he is not going to return to his state.' The T'ai Kung, overhearing it, got dressed that night and set out, reaching his state just before first light. The Marquis of Lai came out to attack, and fought with him for Ying-ch'iu, which bordered Lai. The people of Lai were Yi people who, taking advantage of the chaos under King Chou and the new settlement of the Chou dynasty, assumed Chou would be unable to assemble the distant quarters. For this reason they battled with the T'ai Kung for his state.

"When the T'ai Kung reached Ch'i he rectified the government in accord with prevailing customs; simplified the Chou's forms of propriety; opened up the occupations of the merchants and artisans; and facilitated the realization of profits from fishing and salt. The people turned their allegiance to Ch'i in large numbers, and Ch'i thus became a great state.

"When the youthful King Ch'eng ascended the Chou throne and the late King Wu's brothers Kuan Shu and Ts'ai Shu revolted, the Yi people in the Huai River valley again turned against the Chou. So King Ch'eng had Duke Chao K'ang issue a mandate to the T'ai Kung: 'To the east as far as the sea, the west to the Yellow River, south to Mu-ling, and north to Wu-ti, thoroughly rectify and order the five marquis and nine earls.' From this Ch'i was able to conduct a campaign of rectification to subdue the rebellious and become a great state. Its capital was Ying-ch'iu.

"Probably when the T'ai Kung died he was more than a hundred years old.

"The Grand Historian comments: 'I went to Ch'i—from Lang-yeh, which belongs to Mount T'ai, north to where it fronts the sea, embracing two thousand kilometers of fertile land. Its people are expansive, and many conceal their knowledge. It is their Heaven-given nature. Taking the T'ai Kung's Sageness in establishing his state, isn't it appropriate that Duke Huan flourished and cultivated good government, and was thereby able to assemble the feudal lords in a covenant. Vast, vast, truly the style of a great state!'"

Despite this detailed biography which even records divergent views of his origins in Ssu-ma Ch'ien's generally reliable *Shih Chi*, over the millennia Confucian skeptics have even denied the T'ai Kung's very existence. Others, perturbed by his apparently lowly background, consigned him to a minor role. Both justified their views by the absence of any references to him or his activities in the archaic texts that were traditionally believed to authentically chronicle these epoch-making events, the *Book of Documents* and the *Spring and Autumn Annals*. Thus, they generally followed the second great Confucian, the pedantic Mencius, in refusing to accept the brutal nature of military campaigns and the inevitable bloodshed that occurred at the battle of Mu-yeh. Moreover, King Wu's herculean efforts over the many years prior to the conquest, and his achievements in imposing rudimentary Chou control over the vast Shang domain, similarly tend to be slighted. Consequently, the two figures historically associated with sagacity, Virtue, and the civil—King Wen and the Duke of Chou—are revered, while the strategist and the final commander, outstanding representatives of the martial, are deprecated or ignored, much in consort with the hypocritical attitudes of the elite but effete Confucian bureaucrats who systematically disparaged the martial and thereby doomed imperial China to weakness and subjugation. However, other scholars and historians, after examining the numerous stories and references to him in disparate texts and winnowing away the legendary and mythic material, have concluded that the T'ai Kung not only existed, but also played a prominent role, much as described in the *Shih Chi's* records. While the details of his initial encounter with King Wen will likely remain unknown, the T'ai Kung was probably a representative of the Chiang clan with whom the Chou was militarily allied and had intermarried for generations. No doubt, as with the Hsia dynasty, whose formerly mythic existence assumes increasingly concrete dimensions with the ongoing discovery of ancient artifacts, the T'ai Kung will eventually be vindicated by historical evidence.

Principles and Concepts

Insofar as the book's compilers deliberately formulated the contents as a series of dialogues between the two Chou kings and the T'ai Kung—the former posing questions or describing situations that require a carefully calculated response, the latter providing both specific tactical answers and broadly-based advice—the confrontation between the Chou and the Shang defines much of the work's orientation and apparent mindset. However, even a cursory examination of the entire work, the longest of *The Seven Military Classics*, immediately reveals it to be the most comprehensive of the early writings, one that subsumes much that preceded it, both philosophically and tactically. In fact, it may well be viewed as the systematized presentation of widely ranging

Warring States strategic concerns and military doctrines, all effectively integrated within a realistic approach heavily influenced by the
Taoist teachings embodied by the famous *Tao Te Ching* and the
humanistic, if naively optimistic, thrust of Mencius and the Confucians. Accordingly, the discussion below—which seeks to provide
some initial orientations for the interested reader—proceeds from a
brief examination of the T'ai Kung's teachings aginst the known historical background before considering them in slightly greater detail
within the context of Taoist and Confucian principles. Throughout, it
should be remembered that *The Six Secret Teachings* apparently
assumed final form toward the end of the Warring States period—
roughly the middle to late third century BCE—about the time the
extant versions of the *Tao Te Ching* fully solidified or about a century
after Sun Pin and Mencius. Therefore, within the context of theoretical discussions, references to the T'ai Kung should be understood as
indicating the speaker visible in *The Six Secret Teachings* rather than
the historical figure of earlier times.

   In order to realize their objective of surviving and then conquering,
the Chou required a grand strategy to develop a substantial material
base, undermine the enemy's strength, and create an administrative
organization that could be imposed effectively in both peace and war.
Accordingly, in concord with most of the military writings, in *The Six
Secret Teachings* the T'ai Kung is a strong proponent of the fundamental Confucian doctrine of the benevolent ruler, with its consequent
administrative emphasis upon the people's welfare. He advocates this
policy in the belief that a well ordered, prosperous, satisfied populace
will both physically and emotionally support their government. Moreover, only a materially sufficient society has the resources to allow
training and instructing the people; the capability to generate the spirit and raise the supplies essential to military campaigns; and the environment to furnish truly motivated soldiers. Furthermore, benevolent
governments immediately become attractive beacons to the oppressed
and dispirited, to refugees, and to other states suffering under the
yoke of despotic powers. They create the confidence that, should a
new regime be established, its rulers will not mimic the errors of
recently deposed evil monarchs.

   The T'ai Kung's basic principles, general policies, and strategic concepts, as expressed in *The Six Secret Teachings*, may be briefly summarized as follows.

CIVIL AFFAIRS

PROFIT THE PEOPLE   The T'ai Kung strongly advocates policies similar to Mencius's historically significant emphasis upon the welfare and condition of the people. Stimulating agriculture must be primary and should encompass positive measures to increase productivity, as well as conscious efforts to avoid interfering with the agricultural seasons, thereby minimizing the negative impact of governmental actions. Only upon an adequate material basis can virtues be inculcated and demands successfully imposed upon the populace. A prosperous, well-governed state inhabited by a contented people will invariably be respected by other powers.

INSTITUTE A STRONG BUREAUCRACY AND IMPOSE CONTROLS Although government must be founded upon moral standards and should assiduously practice virtue, it can only govern effectively by creating and systematically imposing a system of rewards and punishments. These policies must be implemented by a strong bureaucracy composed of talented men carefully selected after insightful evaluation. Values inimical to the state, such as private standards of courage, should be discouraged. However, tolerance must be extended to allies and efforts made to avoid violating their local customs.

Rewards and punishments must be clear, immediate, and universal so that they will become part of the national consciousness. While laws and punishments should be restrained and never multiplied, those few necessary to the state's survival should be rigorously enforced. Punishments should extend to the very highest ranks and rewards down to the lowest. Only then will they prove effective, and people be motivated to observe them irrespective of their positions and whether their potential transgressions might be detected.

PERSONAL EXAMPLE AND SYMPATHY OF THE RULER   The ruler, and by implication all the members of government, should intensively cultivate the universally acknowledged virtues: benevolence, righteousness, loyalty, credibility, sincerity, courage, and wisdom. Since all men love profits, pleasure, and Virtue and detest death, suffering, and evil, the ruler should develop and foster these in common with the people. Ideally he must perceive their needs and desires, and avail himself of every source of information to understand their condition. Personal emotions should never be allowed to interfere with the impartial administration of government, nor should the ruler's pleasures or those of the bureaucracy become excessive, thereby impoverishing the people and depriving them of their livelihood. The ruler should strive to eliminate every vestige of evil in order to forge a persona which dramatically contrasts with the enemy's perversity, vividly presenting the people with diametrically opposed alternatives. Righteousness must always dominate personal emotions and desires, and the ruler should not only actively share both hardships and pleasures with his people, but also project an image of having done so. This will personally bind the people to him, and guarantee their allegiance to the state.

MILITARY AFFAIRS

Much of the book is devoted to detailed tactics for particular situations. However, the T'ai Kung also advised on many topics, including campaign strategy, the selection of generals and officers, training, the preparation and types of weapons, the creation of new weapons, communications, battlefield tactics, and military organization. Although many of his strategies and observations have of course become obsolete, others continue to have enduring value and to be embodied by modern maneuver warfare doctrine. Articulation, segmentation and control, and independent action—not to mention specialized weapons systems and their forces—are all extensively discussed. The following particularly merit summary introduction.

TOTAL WARFARE   One reason *The Six Secret Teachings* was excoriated over the centuries is that the T'ai Kung depicted in the text vociferously insisted upon utilizing every available method to achieve victory, just as the historical T'ai Kung apparently did during the Chou's effort to conquer the Shang. Important measures include always anticipating the possibility of hostilities by consciously planning to employ the normal means of production for warfare; feigning and dissembling to deceive the enemy and allay suspicions; using bribes, gifts, and other methods to induce disloyalty among enemy officials and cause chaos and consternation in their ranks; and further increasing their profligacy and debilitation by furnishing the tools for self-destruction, such as music, wine, women, and fascinating rarities. Spies should be employed while complete secrecy is mandated, and when the battle is joined no constraints imposed.

THE GENERAL   The general must be carefully selected and properly invested in his role as commander-in-chief with a formal ceremony at the state altars, being thereafter entrusted with absolute authority over all military matters. Once he has assumed command the ruler cannot interfere with either his actions or decisions, primarily because valuable opportunities might be lost or actions forced which endanger the army, but also to prevent any of the officers from questioning the general's authority by presuming on their familiarity with the king.

Generals and commanders should embody critical characteristics in balanced combinations to qualify them for leadership, while always being free from traits that might either lead to judgmental errors or be exploitable and thereby doom their forces. Several insightful chapters enumerate these essential aspects of character and their correlated flaws, and suggest psychological techniques for evaluating and selecting military leaders.

ORGANIZATION AND UNITY   Both the military and civilian spheres must be marked by unity and thorough integration if they are to be effective. Individual sections must be assigned single tasks, and an integrated system of reporting and responsibility implemented. A command hierarchy must be created and imposed, fully staffed not only by general officers, but also by technical and administrative specialists.

BATTLE TACTICS   In *The Six Secret Teachings* the T'ai Kung analyzes numerous battle situations and formulates some general principles to guide the commander's actions and efforts to determine appropriate tactics based upon objective classifications of terrain, aspects of the enemy, and relative strength of the confrontational forces. Basically there are two categories: one in which the army is about to engage an enemy, and one in which it suddenly finds itself at a disadvantage in a forced encounter. The topics covered include selection of advantageous terrain; assault methods against fortifications; night attacks; counterattacks; escape from entrapment; forest warfare; water conflict; mountain fighting; valley defense; survival under incendiary attack; situations and topography to avoid; techniques for psychological warfare; probing and manipulating the enemy; ways to induce fear; and methods for deception.

Attempts to objectively classify battlefield situations, analyze the enemy, and predict the outcome of engagements assuredly began with the rise of organized combat in antiquity. The first recorded systematic efforts are found in Sun-tzu's characterizations preserved in *The Art of War*, but the other *Seven Military Classics* also contain similar situational descriptions and tactical suggestions. However, those incorporated in *The Six Secret Teachings* are not only far more extensive and detailed, but also differ fundamentally in reflecting the complexities of large-scale mobile and siege warfare. Particularly noteworthy are the exposition and application of separate principles for the three types of component forces—cavalry, infantry, and chariot—depending upon the terrain, battlefield conditions, and composition of the enemy. Mobility and the use of unorthodox tactics are particularly emphasized and probably reflect the results of significant battles that occurred in the fourth and third centuries BCE wherein smaller numbers and weaker forces decisively defeated superior opposition.

Despite the passage of millennia, certain prominent principles, strategies, and tactics from *The Six Secret Teachings* still retain validity and continue to be employed in both the military and business spheres. Clearly the most important are deception and surprise. To maximize an attack's effectiveness, unorthodox measures should be implemented to psychologically and physically manipulate the enemy. Several techniques are possible, but among the most effective are false attacks, feints, and limited encounters designed to constantly harry deployed forces. Thereafter the main attack can be launched, taking advantage of the enemy's surprise and his expectation that it will amount to nothing more than another ruse.

Additional tactics include inciting confusion in the enemy's ranks —for example, through disinformation, then exploiting the ensuing chaos; overawing the enemy through massive displays of force; being aggressive, never yielding the initiative; stressing speed and swiftness; availing oneself of climatic and terrain conditions which trouble and annoy the enemy, such as rain and wind; attacking from out of the sun or at sunset; and mounting intensive efforts to gather intelligence. In accord with the analytic thrust of Chinese military science, the enemy must be carefully evaluated and judgments weighed before a decision whether to attack or defend can be calculated. Weaknesses in an opposing general should be fully exploited, and assaults should be directed toward undefended positions. Traps and ambushes must be avoided, but should always be deployed when in difficulty. Forces should normally be consolidated for effective concentration of power rather than dispersed and weakened. Those who surrender should be spared, to encourage the enemy to abandon their resistance. The troops should be flexible and mobile and their specializations fully utilized. No general should ever suffer a defeat from lack of training or preparation.

THE SIX T'AO

Most commentators characterize the first two Secret Teachings as focusing upon grand strategy and war planning, and the last four as falling within the category of tactical studies. However, insofar as the original authors of *The Six Secret Teachings* either failed to provide any explanations for their apparently thematic groupings, or such prefatory material has been lost, it is difficult to perceive any intrinsic connection between titles such as "dragon" and the contents of the Dragon Secret Teachings. Only the first two Secret Teachings, the Civil and the Martial, which focus upon the twin foundations for conducting warfare—an economically sound, well-administered state with a motivated populace, and a strong army—justify their titles.

Even though a few attempts have been made to discern thematic issues underlying the six individual classifications, such distinctions often appear inadequate to support assigning a particular chapter to one Teaching or another without knowledge of the extant work. Although the table of contents provides a general indication of each Teaching's topics and the translator's introduction surveys the main subjects in some detail, a brief characterization of the individual Teachings may still be useful.

CIVIL T'AO   Moral, effective government is the basis for survival and the foundation for warfare. The state must thrive economically while limiting expenditures; foster appropriate values and behavior among the populace; implement rewards and punishments; employ the Worthy; and refrain from disturbing or harming the people.

MARTIAL T'AO   The Martial Secret Teaching continues the Civil T'ao's discussion of political rather than military measures. It commences with the T'ai Kung's analysis of their contemporary political world and his assessment of their prospects for successfully revolting against the Shang if their avowed objective is to save the world from tyranny and suffering. Attracting the disaffected weakens the enemy and strengthens the state; employing subterfuge and psychological techniques allows for manipulating the enemy and hastening their demise. The ruler must visibly cultivate his Virtue and embrace governmental policies that will allow the state to compete for the minds and hearts of the people, and thus realize Sun-tzu's ideal of victory without engaging in battle.

DRAGON T'AO   The Dragon Secret Teaching primarily focuses upon military organization, including the specialized responsibilities of command staff; the characteristics and qualifications of generals, and methods for their evaluation and selection; the ceremony appropriate for commissioning a commanding general, to ensure that his independence and awesomeness are established; the importance of rewards and punishments in creating and maintaining the general's awesomeness and authority; and essential behavior if the general is to truly command in person and foster allegiance and unity in his troops.

Secondary issues concern military communications and the paramount need for secrecy; evaluating the situation and acting decisively when the moment arrives; understanding basic tactical principles, including flexibility and the unorthodox, while avoiding common errors of command; various indications and cues for fathoming the enemy's situation; and the everyday basis for military skills and equipment.

TIGER T'AO   The Tiger Secret Teaching opens with a discussion of the important categories of military equipment and weapons, but then continues with widely ranging expositions on tactical principles and essential issues of command. Although types of deployment are briefly considered, and the necessary preparation of amphibious equipment addressed, most of the chapters advance tactics for extricating oneself from adverse battlefield situations. The solutions provided generally emphasize speed, maneuverability, unified action, decisive commitment, the employment of misdirection, the establishment of ambushes, and the appropriate utilization of different types of forces.

LEOPARD T'AO   The Leopard Secret Teaching emphasizes tactical solutions for particularly difficult types of terrain, such as forests, mountains, ravines and defiles, lakes and rivers, deep valleys, and other constricted locations. It also discusses methods to contain rampaging invaders, confront superior forces, deploy effectively, and act explosively.

CANINE T'AO   The most important chapters in the Canine Secret Teaching expound detailed principles for appropriately employing the three component forces—chariots, infantry, and cavalry—in a wide variety of concrete tactical situations, and discuss their comparative battlefield effectiveness. Another section describes deficiencies and weaknesses in the enemy which can and should be immediately exploited with a determined attack. Finally, several chapters address general issues that seem more appropriate to the Dragon Secret Teaching: the identification and selection of highly motivated, physically talented individuals for elite infantry units, and for the cavalry and chariots; and methods for training the soldiers.

Taoism and The Six Secret Teachings

Although considerations of space preclude a thorough discussion of the nature and development of Taoist military thought, the prevalence of Taoist conceptions in the fervent intellectual life of the turbulent Warring States period and their readily discernible incorporation in *The Six Secret Teachings* merit at least brief consideration. Accordingly, while several important issues will be systematically explored, rather than striving for comprehensiveness, the following analyses are confined to outlining a few interesting conclusions and elucidating some impressions on selected topics, essentially suggestions for contemplation. Translations of critical *Tao Te Ching* chapters will be provided; however, they differ somewhat from traditional interpretations because our orientation emphasizes military concepts and explicating associated, but frequently ignored, implications. Interested readers are encouraged to consult any of the several standard translations produced over the past decades, such as Waley's *The Way and Its Power*, Wing-tsit Chan's erudite *The Way of Lao-tzu*, or D. C. Lau's *Tao Te Ching*, for comparison.

The viewpoints and beliefs subsumed by the famous Taoist classic known as the *Tao Te Ching* not only provided the critical impulse for what became known under the general rubric of Taoism, but also the main thrust for many strains of thought within the military writings as well. (The other major early Taoist book, the *Chuang-tzu*, had no immediate effect on pre-Imperial military theorists, including the late *Three Strategies of Huang Shih-kung*, which was authored sometime late in the Warring States period.) Attributed to one "Lao-tzu," a cognomen meaning something like "Old One," the *Tao Te Ching* was traditionally considered a late Spring and Autumn period work based upon a story, preserved in the *Shih Chi*, that Confucius himself visited Lao-tzu in his search for wisdom. Although the story is entertaining, most present-day scholars discard such apocrypha in favor of textual and historical analyses that conclude that the extant *Tao Te Ching* is a loosely integrated compilation of variegated sayings which probably assumed present form about the end of the third century, just when *The Six Secret Teachings* was being finalized and theories of yin-yang and the four seasons systematically correlated with both natural and prognosticative phenomena. However, since the *Tao Te Ching* certainly evolved over a century or two, some of its core verses may date back even to the Spring and Autumn period, as well as to several different locales. This evolutionary origin well accounts for the presence of disparate threads of thought—cursorily joined due to apparently common themes, phrases, or even single words—not only throughout the book, but also in the individual chapters themselves.

Lao-tzu's biography in the generally reputable *Shih Chi* identifies him as a native of Ch'u, a southern, peripheral state known as the source of many esoteric wonderings and wanderings. Remarkably, one tradition claims that Lao-tzu's own son subsequently became a distinguished general in the powerful state of Wei early in the Warring States period, suggesting that the dichotomy between military affairs and Taoist thought was less extensive than normally claimed. However, little attention has ever been accorded the conceptual material and tactical principles devoted to warfare embedded in the *Tao Te Ching*, whether simply as tangential issues or as essential concerns of government. Several chapters clearly focus on military affairs and others raise relevant aspects, leading General Wei Ju-lin, for example, to view Lao-tzu as the progenitor of psychological warfare, because he stressed spirit over material substance; mental warfare, because he emphasized thought over military weapons; and plans and stratagems, because he stressed techniques of the Tao over the tactical imbalance of power. However, many of the *Tao Te Ching*'s perceptions have much in common with the contents of the early military classics, including *The Six Secret Teachings* and *Three Strategies of Huang Shih-kung*—not just specific concepts and general principles, such as may be found even in *The Art of War*, which probably dates from very early in the Warring States period, but especially the vision and mindset. In fact, the first two *t'ao* or secret teachings of *The Six Secret Teachings* may be understood as expositions very much in the tradition of the *Tao Te Ching*, while the second half of the book focuses on fundamental military principles and concrete tactics that often embrace and exploit Taoist conceptual approaches. Consequently, while Lao-tzu clearly pondered military issues, some traditional thinkers equally view the T'ai Kung portrayed in *The Six Secret Teachings* as the progenitor of Taoist military concepts and strategies, as well as the father of tactics in China.

Two Essential Taoist Critiques

The *Tao Te Ching* emphasizes two deprecatory beliefs: governments are generally inimical to the people, particularly when they actively engage in the task or enterprise of governing, but especially when they become corrupt, repressive, and extravagant; and among the state's activities, warfare is the most heinous. Although strongly condemnatory, neither assertion completely precludes the Sage from undertaking properly constrained and harmonized efforts in both spheres, and in fact the *Tao Te Ching* outlines an extensive program for governing the state and preserving it from external enemies. However, throughout history most readers have been disinclined to accept the latter belief in favor of grudgingly focusing on the former. The book's strident condemnations, even though probably familiar, bear briefly retelling.

With the gradual erosion of Chou authority and the development of distinct power centers among the more significant, evolving states, China entered into the unremitting misery of the Spring and Autumn and Warring States periods. Perhaps at least partly in emulation of the incredible opulence that had marked the Shang and Chou ruling houses, the individual states vigorously competed to foster their economies, augment their power and territory, and contend for glory under Heaven's dour gaze. This prompted the authors of the *Tao Te Ching* to decry the trend toward wealth being concentrated among the ruling elite at the expense of the peasants:

> The court is excessively contaminated,
> While the fields are extremely weedy,
> And the granaries are very empty.
> They wear colorful, embroidered clothes,
> Bear sharp weapons,
> Surfeit themselves with drink and food,
> And have wealth and goods in surplus.
> This is termed robbery,
> It is not the Tao!

Chapter 53 thus clearly disparages an era of decadence in which warfare already troubles the land, when otherwise productive fields have become barren wastelands, and wealth wastefully concentrated in the hands of a few.

An incisive portrait of debauched, aggressively activist government appears in chapter 75:

> The people are famished because their superiors consume
>   excessive taxes, making them hungry.
> The people are difficult to administer because their superiors
>   are active, making them difficult to govern.
> The people are untroubled by death because their superiors
>   seek life too extensively, making them oblivious to death.
> Now it is only one who does not act for life that is morally
>   superior to one who values life.

Two consecutive chapters in the *Tao Te Ching*—30, immediately below, and 31, following—vehemently condemn warfare:

> One who assists the ruler in accord with the Tao does not
>   coerce the realm with weapons, for such affairs easily
>   rebound.
> Wherever the army has encamped, thorny brambles will
>   grow.
> After large armies have flourished there will certainly be
>   baleful years.
> One who excels rests in the results, that's all,
> But does not thus dare to maintain great strength.
> He rests in the result without bragging,
> He rests in the result without boasting,
> He rests in the result without becoming arrogant.
> He rests in the result because he has no alternative,
> He rests in the result but does not manifest his might.
> Things that are strong grow old, the old is contrary to the Tao.
> What is contrary to the Tao perishes early.

Chapter 30 thus asserts that the raising and employing of troops invariably entails disaster, both for one's own state and for all the territories that suffer the army's passage. Succinctly, how could warfare bode well for anyone in antiquity when, apart from the rampant destruction incurred on actual battlefields, foraging and plundering would denude the land; men and animals foul the water; and encampments and battles destroy the earthworks laboriously constructed over generations and centuries? The army could hardly be regarded as other than a truly "inauspicious implement," for its impending arrival clearly signaled imminent doom and destruction, if not vanquishment and extinction.

Accordingly, the first sentences of chapter 31 in the tomb text continue:

> Now weapons are inauspicious implements,
> There are things that abhor them.
> Thus one who has attained the Tao does not dwell among
> them.

These verses certainly imply that military affairs are to be totally shunned and, in accord with the tenor of much of the book, a pacifist or quietist stance adopted. Furthermore, the first line might equally well be translated as "armies are inauspicious implements," for the term *ping*, while originally meaning "weapons," by extension eventually referred to the army as a whole. Armies being thus brutally condemned, chapter 36 concludes that they (or weapons) should be concealed, admonishing that "the state's sharp implements cannot be displayed to the people."

Although chapter 31 appears to have commentary intermixed with the text, the subsequent lines remain provocative:

> The perfected man honors the left in normal affairs,
> But when employing the army he honors the right.
> Armies are inauspicious implements,
> Not the instruments of the perfected man,
> But when he has no alternative but to employ them,
> He emphasizes calmness and equanimity.
> Achieving victory, he does not glorify it;
> For glorifying it is to take pleasure in killing men.

One who takes pleasure in killing men will never achieve his
  ambition under Heaven.
Auspicious affairs stress the left,
Inauspicious affairs the right.
Subordinate generals occupy the left,
The commanding general the right.
This states that one manages them as if in rites of mourning.
When one kills masses of men, he should weep for them
  with grief and sorrow.
When one is victorious in battle, he should treat the situation
  with rites of mourning.

In this single chapter a dichotomization in Taoist thought may be
clearly seen, for warfare—just as in Mencius's view which severely
condemned it—still retains a certain applicability, still remains
inescapably necessary under some circumstances. However, in conso-
nance with the overall vision and tone of the *Tao Te Ching*, even this
realistic but minimal admissibility is tempered by the realization that
warfare is ultimately an inhumane and sorrowful affair. Therefore,
because men have perished—both the enemy's and one's own—victories should prompt grief and weeping rather than elicit joy and
jubilation. Moreover, while this chapter has often been characterized
simply as a condemnation of warfare, it is neither an absolute nor an
overwhelming one. Fundamentally, it is aggressive warfare that
deserves true abhorrence; defensive warfare continues to be a viable,
though sorrowful, option. So viewed, the prosecution of military
affairs is far more circumscribed than even Mencius—who wished
to have all experts in the military arts executed as criminals—would
propose.

FUNDAMENTALLY DEFENSIVE AND
REACTIVE NATURE OF TAOIST WARFARE

Within the framework of permissible confrontations, the *Tao Te Ching*
provides a clear vision of warfare as defensive and reactive rather than
aggressive and proactive. The most remarkable exposition of this view
appears in chapter 69, which commences with an exposition of the

concepts of "guest" and "host," a fundamental distinction found in several military writings, especially Sun Pin's *Military Methods* and the later *Questions and Replies*, although not explicitly in *The Six Secret Teachings* (further illustrating that *The Six Secret Teachings* is a distinct work, with its own internal organizing vision rather than simply an eclectic compilation of common materials). In normal military usage, "guest" generally refers to an invader, "host" to a defender fighting in his home state or on terrain he already occupies. However, chapter 69 of the *Tao Te Ching* opens with a somewhat puzzling passage that inverts the designations:

> Among those who employ the military there is a saying which runs:
> "I do not dare to act as the host but act as the guest; I do not dare
> to advance an inch but withdraw a foot."

Based upon commentators' explications over the millennia, it must be concluded that Lao-tzu is employing the terms "guest" and "host" in an idiosyncratic fashion because "host" here refers to the party initiating the combat—the "master" of the situation—and the "guest" to the party that responds. In this light it becomes apparent that he is advocating the adoption of what might best be termed a temporizing strategy, one designed to force the enemy to commit and largely expend their energies before responding in detail. This tactic, one of several possible, appears prominently in the military writings in general, as well as some parts of *The Six Secret Teachings*. (The text devotes several chapters in the Tiger and Leopard Secret Teachings to the tactics for invading forces and for countering responses.) However, Lao-tzu's approach is not simply philosophical, reflecting the Taoist emphasis upon yielding and deference to overcome the brutal and powerful (as will be discussed below), but inherently accords with the quietist response to a perverse environmental context. Survival, being almost insurmountably difficult, requires a certain ruthless commitment to such measures, to continued yielding, as evidenced by chapter 73 of the *Tao Te Ching*:

> One who is courageous in daring will be killed.
> One who is courageous in not daring will live.

Moreover, from the Spring and Autumn period into the early Warring States period, when the *Tao Te Ching*'s concepts were perhaps taking form, the defense enjoyed not just the historically attested threefold advantage over aggressor forces, particularly if the battleground had been carefully chosen and improved, but a virtually invincible position when ensconced in even minimal fortifications. While sieges eventually became critical operations during the middle Warring States period as cities became strongpoints and important targets that could no longer be ignored (prompting, in turn, the rapid improvement of siegecraft), any battlefield measures a defender might take, including withdrawing to force the aggressor to expend his limited supplies and energies on unfamiliar, hostile ground, would considerably augment the advantage. Consequently, tacticians from Sun-tzu onward generally adopted the principle of manipulating and destabilizing the enemy, wearying and exhausting him, depleting his spirit and material, before engaging in actual combat, in order to wrest victory with the least expenditure of men and equipment.

The ultimate objective of such measures, the very pinnacle of strategy, would be emerging victorious without ever engaging the enemy in physical combat. An ideal first advanced in Sun-tzu's *Art of War*, it is equally visible in *The Six Secret Teachings*—particularly chapter 13— and the *Tao Te Ching* whose chapter 68 states:

> One who excels as a warrior is not martial.
> One who excels in combat does not get angry.
> One who excels in conquering the enemy does not engage in
>   battle with him.
> One who excels in employing men acts deferentially to them.
> This is what is termed the Virtue of nonconflict.
> This is what is termed strength in employing men.
> This is what is termed matching the extremity of Heaven.

However, while embracing this transcendent vision of victory, the chapter surprisingly contradicts the fundamental military belief that anger proves necessary for soldiers to kill their counterparts. Even though the military theorists believed anger would adversely affect the general's judgment, it was strongly felt that the infantry could not suc-

cessfully overcome the dangers and difficulties of the battlefield without rage and emotional fervency. Therefore elaborate oaths and stimulating speeches that vilified the enemy were invariably employed to
stir up the men immediately before battle. As Sun-tzu said, "What
stimulates the men to slay the enemy is anger."

The *Tao Te Ching*, conceived when individual existence was daily
threatened and states were constantly perishing, also emphasizes
the "victory" of mere survival that might be attained by simply not
fighting, by adamantly refusing to "compete" in a contest of violence.
Lao-tzu envisioned the paradigm of noncompetition (or, more evoca-
tively, "nonconflict") in the operation of Heaven, mimicked by the
Sage with whom, "because he does not compete, none under Heaven
are able to compete." Chapter 81 similarly asserts that "The Tao of
Heaven is not to compete [with things], but yet it excels at achieving
victory." Of course such references to Sage rulers—who deliberately
eschew competition in order to become first, and thus engage in a
sort of teleological subterfuge against the Tao—and Heaven, whose
strategic power is inestimable and unstoppable, and therefore need
not compete (but underlies the natural cycle of *yin* and *yang*, and
thus death and destruction when *yin* ascends) have little relevance for
individuals or even states beset by armed enemies. Accordingly, the
military writers, most government administrators, and many philoso-
phers, especially those identified with the "realist" or "Legalist" school,
found Lao-tzu's essential pacifism to be fundamentally deficient, if
only because of life's temporal nature. In short, the failure of the
quietist approach—which might save isolated individuals who, like
Yang Chu, successfully escaped into China's inaccessible mountains
to forever eke out a hermitlike survival—is that weakly submitting
is just as likely to see states destroyed and individuals killed as permit-
ted to live on under their conquerors. Although every strong, brutal
enemy will inevitably perish, whether from the inescapable tendency
of Heaven to repress the excessive, the ineffable movement of the
Tao toward reversing fulminating strength and over-extension, or
simply old age, the time frame remains problematic. In the eyes of
the military writers, deliberate nonactivity, defined as the ruler being
neither rapacious nor overly involving the government in a myriad
of adverse affairs, represents the political ideal, but only if his ad-

ministration equally emphasizes national prosperity and subtle state-building. In this view, expressed throughout *The Six Secret Teachings*, the purely quietist approach fails to embrace the policies essential for promoting the people's welfare, harmonizing with the seasons, eliminating harm and perversity, and defending the state. The battle-field situation provides a paradigm: Lao-tzu's view might well save individuals who strive mightily to minimize their exposure to risk, but the military writers observed that those who cast away their lives save them.

Although this sort of battlefield abandonment coheres with the Taoist spirit of not clinging to life and thereby regaining it, found more in *Chuang-tzu* and later expressions than in the *Tao Te Ching*, chapter 50 observes that while striving after life leads to death, perfection in the Tao also ensures a degree of invincibility:

> Man comes forth into the realm of life and ends by entering
>   the abode of death.
> Three out of ten are partisans of life,
> Three out of ten partisans of death,
> And among the living, those that tend toward deadly terrain,
>   also three out of ten.
> Now how is this so ?
> Because they employ their lives to live too intensively.
> You have probably heard that one who takes hold of life,
> When traveling about the land does not encounter rhinos or
>   tigers,
> Nor when he enters the army suffers weapons or armor.
> Buffalo have nowhere to impale their horns,
> Nor tigers anywhere to sink their claws,
> Nor weapons to engage their blades.
> For what reason ?
> Because he has no deadly terrain within him.

Of course, this merely shifts the difficulty to realizing the Tao rather than escaping from physical harm through decisive action and fortu-nate coincidence.

TAOIST VISION OF RULERSHIP

The antidote to human misery and suffering, to warfare and its consequences, is a Sage serving as ruler, someone who recognizes the nature of phenomenal progressions and harmonizes with their cyclic nature, governing through studied rather than ignorant or naive inactivity. As this view is well known, it only remains to note some of the more essential passages in the *Tao Te Ching*. Perhaps the most famous, one that incidentally introduces the critical concept of the unorthodox, is chapter 57 of the traditional recension:

> With the orthodox govern the state,
> With the unorthodox employ the military,
> And without mounting affairs take All under Heaven.
> How do I know that this is so ?
> As follows:
> As the prohibitions and interdictions under Heaven are
>   multiplied,
> The people grow increasingly impoverished.
> As the number of sharp implements in court multiplies,
> The state will be increasingly muddled.
> As human skill and artifice multiply,
> The more rarities will be brought forth.
> As laws and edicts are increasingly publicized,
> The more thieves and robbers there will be.
> Thus the Sage says: I am inactive and the people are
>   transformed of themselves,
> I love tranquillity and the people become orthodox [upright]
>   by themselves.
> I mount no affairs and the people become rich by
>   themselves.
> I am without desires and the people become simplified by
>   themselves.

Appropriately, when this vision of rulership under a Taoist Sage is attained, good order is realized, the people prosper in the essentials of life, harmony prevails, and military affairs are thereby precluded. As depicted by chapter 46:

> When the Tao prevails throughout the realm,
> Swift steeds are released to fertilize the fields.
> When the Tao does not prevail throughout the realm,
> War horses multiply amidst the suburbs.
> No misfortune is greater than not knowing the point of
>   sufficiency,
> No calamity greater than being covetous.
> Therefore the contentment of knowing sufficiency is
>   everlasting contentment.

Unrest obviously requires the employment of military forces and equipment; war horses are therefore valued. (It might be noted that the desires themselves are being condemned here, whereas other Taoists, including Chuang-tzu and the later Neo-Taoists, would advise experiencing them rather than completely denying them, ever maintaining an inner core of tranquillity that will naturally prevent their effects from becoming excessive.)

Although war is never specifically condemned, and in fact must be undertaken for humanitarian purposes, much of this grand Taoist vision (as well as some other significant concepts) underlies and pervades *The Six Secret Teachings*, especially in the essential early chapters found in the Civil and Martial Secret Teachings. In particular, "Preserving the State," "Opening Instructions," and "Instructions on According with the People" in orientation, tone, and content remarkably unfold essentially Taoist programs and concerns, even though seasonal theory plays a more prominent role in the T'ai Kung's discussion than in the *Tao Te Ching*. The following pivotal passage from "Preserving the State" illustrates his approach, including the emphasis on not being willfully assertive:

> Heaven gives birth to the four seasons, Earth produces the myriad
> things. Under Heaven there are the people, and the Sage acts as
> their shepherd. Thus the Tao of spring is birth and the myriad
> things begin to flourish. The Tao of summer is growth, the myriad

things mature. The Tao of autumn is gathering, the myriad things are full. The Tao of winter is storing away, the myriad things are still. When they are full they are stored away; after they are stored away they again revive. No one knows where it ends, no one knows where it begins. The Sage accords with it, and models himself on Heaven and Earth. Thus when the realm is well ordered, his benevolence and sagacity are hidden. When All under Heaven are in turbulence, his benevolence and sagacity flourish. This is the true Tao.

However, overall *The Six Secret Teachings*—especially the chapters in the Martial Secret Teaching that stress the Sage's transforming influence—not only accepts but also advocates the necessity for certain activist measures of government: encouraging agriculture, promoting the state's prosperity, ensuring civil order, benefiting the people, eliminating harm, attracting immigrants, preparing for the defense of the state and, however unwillingly, undertaking external campaigns to vanquish the perverse and ensure peace under Heaven, for, as Sun-tzu said and the T'ai Kung reiterates, "warfare is the greatest affair of state."

CONCEPT OF TIMELINESS

The bitter reality of military experience convinced the ancient strategists that timeliness is crucial to wresting victory over the enemy. Although much in accord with the paradigm of the four seasons and the lessons easily derived therefrom—such as not planting crops too early in the spring, when they might perish from frost and the ground's residual coldness, or too late, when a fall cold spell may strike before maturation—the tactical concept of timeliness is nearly identical with that embraced by the Taoists, particularly Chuang-tzu, for whom the subtle or incipient moment (termed *chi*) was critical. General impressions to the contrary, whenever assertive action is to be required, such as the application of military force or other measures to stem evil, relieve suffering, and cope with difficulties, the *Tao Te Ching* similarly encourages early action, striking before the situation has unfolded and difficulties become entrenched. Two chapters strongly advance this

position, 63 (which begins with "act without acting") and 64, whose
key sentences unfold as follows:

> Plot against the difficult while it remains easy,
> Act upon the great while it is still minute.
> Difficult affairs throughout the realm invariably commence
>   with the easy;
> Great affairs throughout the realm inevitably commence with
>   the small.
> For this reason the Sage never acts against the great and is
>   thus able to complete greatness.
> . . . .
> What is tranquil remains easily grasped,
> What has not yet betrayed signs is easy to plot against.
> The brittle is easily split,
> The minute is easily scattered.
> Act upon them before they attain being,
> Control them before they become chaotic.
> Trees that require both arms to embrace
> Are born from insignificant saplings.
> A nine-story tower commences with a little accumulated
>   earth.
> A journey of a thousand kilometers begins beneath one's feet.

Accordingly, it is critical to start early, to discern patterns in the
formless, and extinguish problems at their inception. However, difficulties arise with the latter, for few but the Sage can actually discern
matters before they have assumed concrete shape, before they have
manifested themselves and their implications can be fathomed. (Sun-
tzu, of course, exploited human incapabilities in this regard, advocating that the army should be formless, to prevent others from anticipating actions and successfully plotting against them.) Yet an activist
principle still remains, one fully drawn out by *The Six Secret Teachings*.

Within the context of his vision of the ruler—such as is explicated
in "Preserving the State"—while still harmonizing with the natural
phenomena unfolding about him, in several chapters the T'ai Kung

emphasized taking action when necessary. For example, in a passage
that clearly resounds with images from chapter 64 of the *Tao Te Ching*,
he said:

> When the sun is at midday you must dry things. If you grasp a
> knife you must cut. If you hold an ax you must attack. If, at the
> height of day, you do not dry things in the sun, this is termed losing
> the time. If you grasp a knife but do not cut anything, you will lose
> the moment for profits. If you hold an ax but do not attack, then
> bandits will come.
>
> If trickling streams are not blocked, they will become great
> rivers. If you do not extinguish the smallest flames, what will you
> do about a great conflagration? If you do not eliminate the two-leaf
> sapling, how will you use your ax when the tree has grown?

Centuries earlier the great general Wu Ch'i had said : "If the general
is not quick-witted and acute, the Three Armies will lose the moment," while the T'ai Kung concluded:

> One who excels in warfare will not lose an advantage when he per-
> ceives it or be doubtful when he meets the moment. One who
> loses an advantage or lags behind the time for action will, on the
> contrary, suffer from disaster. Thus the wise follow the time and do
> not lose an advantage; the skillful are decisive and have no doubts.

Within the purview of good government and effective military
action, in "The Army's Strategic Power" the T'ai Kung summarized:

> Thus one who excels in warfare does not await the deployment of
> forces. One who excels at eliminating the misfortunes of the peo-
> ple manages them before they appear. Conquering the enemy
> means being victorious over the formless. The superior fighter
> does not engage in battle. Thus one who fights and attains victory
> in front of naked blades is not a good general. One who makes
> preparations after the battle has been lost is not a Superior Sage!
> One whose wisdom is the same as the masses is not a general for
> the state. One whose skill is the same as the masses is not a State
> Artisan.

Consequently, initiating action at the precise moment is so critical that any enemy force that fails to recognize and exploit such opportunities immediately becomes just as vulnerable as if it suffered from other fundamentally disabling conditions, just as Wu Ch'i observed about an indecisive enemy in "Evaluating the Enemy": "When they have lost the critical moment and not followed up on opportunities, they can be attacked." When integrated with the concept of hardness and softness discussed below, a new definition for the activities subsumed by the Tao results, a definition found not only in *The Six Secret Teachings*, but also in *The Lost Books of the Chou*—purportedly a work from the Early Chou period, but probably dating from the late fourth century or the third century—and Sun Pin's *Military Methods*, showing the common threads underlying many of the military and philosophical writings and their virtually universal emphasis upon the importance of timeliness:

> If one sees good but is dilatory, if the time for action arrives and one is doubtful, if you know something is wrong but you sanction it—it is in these three that the Tao stops. If one is soft and quiet, dignified and respectful, strong yet genial, tolerant yet hard—it is in these four that the Tao begins. Accordingly, when righteousness overcomes desire one will flourish; when desire overcomes righteousness one will perish. When respect overcomes dilatoriness it is auspicious; when dilatoriness overcomes respect one is destroyed.

The military strategists also emphasized thoroughness in the effort, never becoming negligent at the conclusion, at the "final moment," to avoid misfortune's resurgence, just as chapter 64 of the *Tao Te Ching* warned: "When the people engage in affairs, just when on the brink of success, they thwart themselves. If one is as solicitous about the end as the beginning, then there will not be any thwarted affairs."

THE SOFT CONQUERS THE HARD

A concept prominently found in the *Tao Te Ching* and several of the military writings—such as Sun Pin's *Military Methods*, the *Three Strategies of Huang Shih-kung*, and the *Wei Liao-tzu*, but only implicitly in *The Six Secret Teachings*—is the inherently dynamic concept of the hard and the soft, the firm and the pliable (or flexible). During the late Warring States period the *Tao Te Ching* vigorously advanced the concept of the soft and pliable, juxtaposed against normal world views that always expect the hard to dominate the soft and predicate actions accordingly. Portions of chapters 76 and 78 succinctly characterize the relationship:

> Alive man is pliable and weak,
> Dead he is firm and strong.
> Alive the myriad things, grasses, and trees are pliable and
>   fragile;
> Dead they are dry and withered.
> Thus the firm and strong are the disciples of death,
> The pliant and weak are the disciples of life.
> For this reason armies that are strong will not be victorious;
> Trees that are strong will break.
> The strong and great dwell below,
> The pliant and weak dwell above.

> Under Heaven there is nothing more pliant and weak than water, but for attacking the firm and strong nothing surpasses it, nothing can be exchanged for it. The weak being victorious over the strong, the pliant being victorious over the firm—there is not anyone under Heaven who does not know this. Yet no one is able to implement it.

These insights were clearly embraced by portions of the *Wei Liao-tzu*, a text composed at the end of the Warring States period:

> The army that would be victorious is like water. Now water is the softest and weakest of things, but whatever it collides with—such as hills and mounds—will be collapsed by it for no other reason than that its nature is concentrated and its attack is totally committed.

Thus formulated, the concept is more complex than originally expressed in the *Tao Te Ching*, for it recognizes that it is not just "softness" that works the change, but rather the water's focus and endurance—its unremitting pressure over time—that cannot be withstood. Historically, Chinese forces frequently employed water as a weapon, either to inundate general areas occupied by an enemy, or as a battering ram achieved by first blocking a stream or small river and then suddenly destroying the dam. Interestingly, Sun-tzu distinguished the applicability of incendiary and aquatic assaults as follows: "Using fire to aid an attack is enlightened, using water to assist an attack is powerful. Water can be used to sever, but cannot be employed to seize."

In general, the military writers perceived a need to employ each one of the four—the soft, hard, pliant, and firm—appropriately. The *Three Strategies* cites an earlier, no doubt well-known, observation in this regard:

> The *Military Pronouncements* states: "The soft can control the hard, the weak can control the strong." The soft is Virtue. The hard is a brigand. The weak is what the people will help, the strong is what resentment will attack. The soft has situations in which it is established; the hard has situations in which it is applied; the weak has situations in which it is employed; and the strong has situations in which it is augmented. Combine these four and implement them appropriately.

As the concluding sentence from this same chapter notes, the four must be integrated and combined. Perversely adopting a single one will doom the state:

> The *Military Pronouncements* states: "If one can be soft and hard, his state will be increasingly glorious! If one can be weak and strong, his state will be increasingly glorious! If purely soft and purely weak, his state will inevitably decline. If purely hard and purely strong, his state will inevitably be destroyed."

Even Sun-tzu, whose *Art of War* betrays Taoist-like influences but was written somewhat earlier, noted: "Realize the appropriate employment of the hard and soft through patterns of terrain." Finally, Wu Ch'i, who was highly concerned about the problems of command and control, indicated the necessity for any qualified general to embrace such capabilities: "Now the commanding general of the Three Armies should combine both military and civilian abilities. The employment of soldiers requires uniting both hardness and softness."

Similar thoughts may be seen underlying the T'ai Kung's choice of various tactics for different situations coupled with the ruler's (and general's) personality and governance of his forces, for even largely identical situations are treated differently rather than simply advocating a concentrated, straightforward onrush. However, within this thrust of the soft overcoming the hard, *The Six Secret Teachings* tends to emphasize more active measures over temporizing and delaying, especially in urgent circumstances, while recognizing that all four are required. (As previously cited, the T'ai Kung said: " If one is soft and quiet, dignified and respectful, strong yet genial, tolerant yet hard—it is in these four that the Tao begins.") In particular, one of the prominent unifying principles of the disparate tactics suggested in the tactical portions is their flexibility, no doubt modeled after water's defining characteristic. The opposite—strong, rigid, inflexible armies executing predetermined tactical plans—incapable of adapting to the noise of battle, the swirling pace of events as combat unfolds, presages failure. Moreover, if true in days of lesser mobility, similar difficulties may be anticipated for modern, highly mobile armies fighting on computerized battlefields, wherein commanders insist upon real-time control and micromanagement in the execution of well-conceived, but enormously complex and therefore paradoxically rigid battle plans. Accordingly, the *Tao Te Ching* states: "One who preserves his flexibility is called 'Strong.'"

THE UNORTHODOX AND ORTHODOX

The opening lines of chapter 57 of the *Tao Te Ching*, already cited in passing above, fundamentally assert that warfare should be unorthodox—as opposed to governmental measures which are upright, correct, or orthodox—yet thinkers to date have failed to adequately correlate this assertion with the essential characteristic of Chinese warfare—namely, a thoroughgoing commitment to the unorthodox, to maneuver warfare and a multiplicity of techniques apart from direct confrontations. While most translators have understood the first line of chapter 57 as "Govern the state with uprightness, employ the army with unusual tactics," in the context of military parameters we believe the following to be more appropriate:

> With the orthodox govern the state,
> With the unorthodox employ the military,
> And without mounting affairs take All under Heaven.

Apart from the fundamental principle of maneuver warfare—destabilizing and manipulating the enemy to allow the decisive application of strategic power—the concept of unorthodox combat naturally embraces psychological operations and covert measures. Chief among those elucidated by *The Six Secret Teachings* in "Civil Offensive"—a chapter for which the T'ai Kung was much excoriated by hypocritical Confucians in later dynasties—are spreading rumors and disinformation, undermining the enemy's morale and capabilities, undertaking intensive intelligence gathering, and similar efforts directed to achieving victory with minimal exertion, and, if possible, without fighting. (These preliminary efforts ensure that the enemy is already tottering before the conflict is initiated, so their defeat follows as naturally and effortlessly as the changes of the seasons, as a leaf falling from a tree.)

Sun-tzu himself might well be considered the concept's progenitor, at least insofar as extant materials allow attribution, and certainly its first proponent. Although the entire *Art of War* is premised upon and largely reflects the implementation of the unorthodox and orthodox, Sun-tzu primarily advanced the concept of the unorthodox in the chapter entitled "Strategic Military Power." The following paragraphs encapsulate his thought:

> What enables the masses of the Three Armies to invariably withstand the enemy without being defeated are the unorthodox and orthodox.
>
> In general, in battle one engages with the orthodox and gains victory through the unorthodox. Thus one who excels at sending forth the unorthodox is as inexhaustible as Heaven, as unlimited as the Yangtze and Yellow rivers. What reach an end and begin again are the sun and moon. What die and are reborn are the four seasons.

The notes do not exceed five, but the changes of the five notes can never be fully heard. The colors do not exceed five, but the changes of the five colors can never be completely seen. The flavors do not exceed five, but the changes of the five flavors can never be completely tasted. In warfare the strategic configurations of power do not exceed the unorthodox and orthodox, but the changes of the unorthodox and orthodox can never be completely exhausted. The unorthodox and orthodox mutually produce each other, just like an endless cycle. Who can exhaust them?

The conception's essence is that engaging an enemy in battle in conventional, expected ways represents orthodox tactics, while employing unexpected, surprise attacks and movements is unorthodox. Naturally everything depends upon the enemy's assumptions and evaluations in a particular situation: if they expect a flanking attack instead of the usual frontal advance in strength, the former then turns out to be expected and therefore, by definition, "orthodox" rather than "unorthodox." The question is thus reduced to a tactical one: how to create false expectations and how to exploit them. For the former the military writings, including *The Six Secret Teachings*, are replete with techniques for confusing and befuddling the enemy; realizing the latter depends upon misdirection and the army's ability to maneuver, segment and recombine, and effect unexpected speed (for example, through the employment of the cavalry arm, which appeared late in the Warring States period and is extensively discussed in the final chapters of *The Six Secret Teachings*).

The paradigm expression and philosophical culmination of the strategy of the unorthodox and orthodox is found in the final chapter of Sun Pin's *Military Methods*, a work composed in the middle Warring States period, although "The Unorthodox and Orthodox" may be somewhat later. In it Sun Pin not only embraced Sun-tzu's basic concepts, but elucidated and expanded them, further systematizing and advancing them by integrating them with the *Tao Te Ching's* cosmogonic philosophy. Sun Pin's comprehensive formulation transcends the original, and the chapter, which remains the most incisive discussion to be found among any of China's many military writings, bears briefly abstracting here:

When form is employed to respond to form, it is orthodox. When the formless controls the formed, it is unorthodox. That the unorthodox and orthodox are inexhaustible is due to differentiation. Differentiate according to unorthodox techniques, exercise control through the five phases, engage in combat with three forces. Things that are the same are inadequate for conquering each other. Thus employ the different to create the unorthodox. Accordingly, take the quiet to be the unorthodox for movement; ease to be the unorthodox for weariness; satiety to be the unorthodox for hunger; order to be the unorthodox for chaos; and the numerous masses to be the unorthodox for the few.

When action is initiated it becomes the orthodox; what has not yet been initiated is the unorthodox. When the unorthodox is initiated and is not responded to, then it will be victorious. One who has a surplus of the unorthodox will attain surpassing victories.

The cyclic character of natural phenomena underlies the concept of the ever-evolving, ever-changing tactics of the unorthodox and orthodox, as Sun-tzu had earlier asserted. The key is the nature of the visible, of that which has attained form. Once something is visibly formed, it can be described; once describable, characteristics can be appended, predications can be made, and plans can be formulated. In the realm of concrete things—which, for Sun Pin and the other military strategists, includes military deployments or dispositions—any tangible form can be opposed and conquered by another. The normal penchant in the West, perhaps stemming from the Greek tradition, has been to oppose force with force, strength with strength. In contrast, strategists opting for the unorthodox advise against this wasteful and often futile approach, advocating instead consciously determining and employing the complementary position, the latter being identified as the "unorthodox." Therefore, it is only necessary to develop the tactics appropriate to any particular battle, to find the form or disposition among the myriad things whose strength will naturally counter and overwhelm the enemy.

Perhaps with such relationships in mind, Sun-tzu himself, in "Vacuity and Substance," had stressed being formless in order to prevent the enemy from discerning one's intentions and disposition, thereby thwarting the development of effective tactics—whether orthodox or unorthodox—for attacking or defending.

> Thus if I determine the enemy's disposition of forces while I have no perceptible form, I can concentrate my forces while the enemy is fragmented.
>   The pinnacle of military deployment approaches the formless. If it is formless, then even the deepest spy cannot discern it or the wise make plans against it.

This coheres well with the idea of being formless found in chapter 69 of the *Tao Te Ching*, explicating the idea of guest and host: "This refers to campaigning without formations, to seizing without baring one's arms, to grasping without weapons, and hurtling forward without an enemy."

While the principles of the unorthodox, as well as maneuver warfare in general, underlie and are realized throughout the tactical principles advocated throughout *The Six Secret Teachings*, especially the last four *t'ao* that focus on concrete situations and their expedient resolution, two clear statements of the unorthodox are also found. The theme of chapter 27, "The Unorthodox Army," is clearly realized in a remarkable set of tactical principles for general application that stresses, but is not limited to, the unorthodox. Some prominent examples of unorthodox techniques include the following:

> When our two armies, opposing each other, have deployed their armored soldiers and established their battle arrays, releasing some of your troops to create chaos in the ranks is the means by which to fabricate deceptive changes.
>   Marshy depressions and secluded dark areas are the means by which to conceal your appearance.
>   Setting up ingenious ambushes and preparing unorthodox troops, stretching out distant formations to deceive and entice the enemy, are the means by which to destroy the enemy's army and capture its general.
>   Unorthodox technical skills are the means by which to cross deep waters and ford rivers.

Disguising some men as enemy emissaries is the means by which to sever their supply lines.

Forging enemy commands and orders, and wearing the same clothes as the enemy, are the means by which to be prepared for their retreat.

Accordingly, following a lengthy series of similar specific measures, the T'ai Kung concluded:

> One who does not know how to plan for aggressive warfare cannot be spoken with about the enemy. One who cannot divide and move his troops about cannot be spoken with about unorthodox strategies. One who does not have a penetrating understanding of both order and chaos cannot be spoken with about changes.

However, it should be noted that the chapter is framed in terms of realizing the ultimate objective of bringing overwhelming strategic power to bear:

> The ancients who excelled at warfare were not able to wage war above Heaven, nor could they wage war below Earth. Their success and defeat in all cases proceeded from the spiritual employment of strategic power. Those who attained it flourished; those who lost it perished.

In addition, an explicit statement about the importance of the unorthodox appears at the beginning of "The Army's Strategic Power":

> Strategic power is exercised in accord with the enemy's movements. Changes stem from the confrontation between the two armies. Unorthodox and orthodox tactics are produced from the inexhaustible resources of the mind. Thus the greatest affairs are not discussed, and the employment of troops is not spoken about. Moreover, words which discuss ultimate affairs are not worth listening to. The employment of troops is not so definitive as to be visible. They go suddenly, they come suddenly. Only someone who can exercise sole control, without being governed by other men, is a military weapon.

The shift to a sole figure of authority, exercising absolute power, of course marks a distinction from the *Tao Te Ching* and accords with the concepts found in Huang-Lao thought, the realpolitik blending of authority and Taoist cosmogony that flourished at the end of the Warring States period.

TACTICAL EXPLOITATION OF UNSTABLE EXTREMES

Closely correlated to the concepts of maturation being deadly and the pliant easily overcoming the stiff is the idea that all extremes are inherently unstable, that reversal or reversion is the ineluctable movement of the Tao. Chapter 81 of the *Tao Te Ching* points out that Heaven represses what is excessive, while chapter 77 asserts it dramatically with the analogy of a bow:

> The Tao of Heaven is like the stretching of a bow!
> It represses the high and raises up the low;
> It reduces the surplus and supplements the insufficient.
> The Tao of Heaven reduces the excessive and supplements
>   the insufficient;
> The Tao of man is not thus, for men reduce the insufficient
>   to increase what is excessive.
> Who is capable of having a surplus in order to increase All
>   under Heaven?
> Only one who has attained the Tao.
> For this reason the Sage acts without reliance;
> His achievements are complete but he does not dwell in
>   them,
> For he does not want to manifest his Worthiness.

Tao Te Ching 58 even observes that "the orthodox overturns to become the unorthodox, that which excels again becomes aberration." However, chapter 40 provides the most succinct characterization of the Tao's dynamics:

> Reversal is the movement of the Tao,
> Weakness is the Tao's employment.

The myriad things under Heaven are all given birth in being,
But being stems from nonbeing.

Therefore, strength is in constant danger of reverting to weakness, power to impotence—which is, of course, the natural course of all human edifices and constructs over the course of history. Consequently, the wise Taoist (and enlightened military commander) seeks to avoid becoming too strong, to preclude this withering or being slashed away. Whether it remains a feasible possibility in a perverse age when others willingly seize every opening, attack and exploit every weakness, must yet be pondered.

The positive correlate or exploitable aspect of this natural tendency in the universe appears in chapter 36 of the traditional recension of the *Tao Te Ching*, which states:

If you want to reduce something, you must certainly
    stretch it.
If you want to weaken something, you must certainly
    strengthen it.
If you want to abolish something, you must certainly make
    it flourish.
If you want to grasp something, you must certainly give
    it away.
This is referred to as subtle enlightenment.
The pliant and weak will conquer the hard and strong.

The psychological warfare techniques elaborated in the chapter entitled "Three Doubts" in *The Six Secret Teachings* clandestinely seek to effect the downfall of an enemy by augmenting a strength or dominant characteristic until it becomes unstable or reaches the breaking point:

Now in order to attack the strong, you must nurture them to make them even stronger, and increase them to make them even more extensive. What is too strong will certainly break; what is too extended must have deficiencies. Attack the strong through his strength.

Other military writings, such as the *Ssu-ma Fa*—which warned against making solid formations heavier, no doubt because they would become unmanageable and inflexible—and the *Military Methods* (in a lengthy chapter entitled "The Dense and Diffuse"), recognized and embraced this essential Taoist principle. Virtually all of them, including *The Six Secret Teachings*, sought to avoid wasteful headlong assaults, stressing instead maneuver warfare and techniques to destabilize and undermine the enemy's forces.

CONCEPT OF YIN AND YANG

Conceptualizing the phenomenal world in terms of a simple duality—light/dark, active/passive, hard/soft, stiff/pliant—evolved into the school of thought known as *yin* and *yang*. While the ultimate origins are obscure, as an articulated concept it dates back into the Western Chou and flourished as an independent view, as well as being co-opted and amalgamated into the theoretical underpinnings of the *I Ching* and other divinatory works in the Warring States period. Although the concept of duality, of opposites in a dynamic polar tension, thoroughly grounds and underlies the *Tao Te Ching*, the terms *yin* and *yang* appear only once, in chapter 42, where a somewhat mystifying line states that "the ten thousand things turn their backs to *yin* and embrace *yang*." However, the well-elaborated concept of the four seasons and their natural progression—originally another strain of thought that evolved to encompass a myriad of disparate phenomena and facilitate the conceptual codification of much human, especially court, behavior with the ultimate political objective of practicing seasonally harmonious activities—strongly underlies the cyclic patterns of thought in the *Tao Te Ching* and *The Six Secret Teachings*. The two concepts are extensively inter-linked and reinforcing, and further integrated with ideas about five phases or elements that developed earlier in the Warring States period, apparently under Tsou Yen's impulse, and thereafter provided a third fountain of descriptive thought for both the Taoists and the military writers. While *yin* and *yang* are mentioned in the first chapter of *The Art of War* and figure prominently in tactical conceptualization thereafter, in *The Six Secret Teachings* there are three interesting uses of the term, showing the continuity of the tradition and common ground of conceptualization. "Preserving the State," already discussed above for its explication of the cyclic nature of the seasons, concludes with the T'ai Kung's dynamic analysis of situations in which a righteous man, with the support of the people, might take action against a tyrant:

> In his position between Heaven and Earth, what the Sage treasures is substantial and vast. Relying on the constant to view it, the people are at peace. But when the people are agitated it creates impulses. When impulses stir, conflict over gain and loss arises. Thus it is initiated in *yin,* but coalesces in *yang.* If someone ventures to be the first leader, All under Heaven will unite with him.

In analyzing an opponent's behavior, the T'ai Kung is portrayed as speaking to King Wu of his *yin* and *yang* aspects, the manifest and concealed:

> You must look at the Shang King's *yang* aspects, and moreover his *yin* side, for only then will you know his mind. You must look at his external activities, and moreover his internal ones, for only then will you know his thoughts. You must observe those distant from him, and also observe those close to him, for only then will you know his emotions.

Furthermore, in common with many other military writers and perhaps the earliest origins of the concept of *yin* and *yang*—the dark and sunny sides respectively of a mountain—defensive measures for heights were delimited in terms of *yin* and *yang* in chapter 47, "Crow and Cloud Formation in the Mountains":

> Whenever the Three Armies occupy the heights of a mountain they are trapped on high by the enemy. When they hold the land below the mountain they are imprisoned by the forces above them. If you have already occupied the top of the mountain, you must prepare the Crow and Cloud Formation. The Crow and Cloud Formation should be prepared on both the *yin* and *yang* sides of

the mountain. Some will encamp on the *yin* side, others will encamp on the *yang* side. Those that occupy the *yang* side must prepare against attacks from the *yin* side. Those occupying the *yin* side must prepare against attacks from the *yang* side. Those occupying the left side of the mountain must prepare against the right side. Those on the right, against the left. Wherever the enemy can ascend the mountain, your troops should establish external lines. If there are roads passing through the valley, sever them with your war chariots. Set your flags and pennants up high. Be cautious in commanding the Three Armies, do not allow the enemy to know your true situation. This is referred to as a "mountain wall."

The Sage is also seen as being in harmony with Heaven, according with the Tao of *yin* and *yang* in a passage that illustrates the integration of cyclic seasonality and the T'ai Kung's overview:

The Sage takes his signs from the movements of Heaven and Earth; who knows his principles? He accords with the Tao of *yin* and *yang,* and follows their seasonal activity. He follows the cycles of fullness and emptiness of Heaven and Earth, taking them as his constant. All things have life and death in accord with the form of Heaven and Earth.

Motivation

Subsequent to Sun-tzu, who little discussed the issue, most of the military writers emphasized the importance of the people and the need to amply provide for their material welfare. In the early chapters *The Six Secret Teachings* repeatedly raises this theme in the belief that a contented, adequately sustained populace will emotionally respond with the allegiance necessary to provide the state with highly motivated soldiers, that the civil precedes the martial. While this view is somewhat more aggressive, even exploitative, than found in the *Tao Te Ching,* key chapters in the latter also postulate that the ruler's solicitude provides the basis for a formidable, motivated army. For example, chapter 67 states:

All under Heaven term my Tao vast and unreal. Only because
    it is vast is it unreal!
If it were real, it would have long been minute!
I have three treasures which I grasp and preserve.
The first is called solicitude. The second is called frugality.
    The third is called not daring to act to precede the world.
Because of solicitude there is courage; because of frugality,
    expansiveness; because of not daring to precede others in
    the world, the ability to be the leader of implements.
Now if we were to abandon solicitude and courage, abandon
    frugality and expansiveness, abandon being last and first, it
    would be fatal!
Now solicitude yields victory in warfare and solidity in
    defense.
When Heaven is about to rescue him, it will protect him with
    solicitude.

Clearly solicitude, otherwise termed benevolence or humanity when
spoken of by the Confucians and military theorists, equates with compassionate government, the very same compassion that sorrows the
ruler when men perish on the battlefield, as previously discussed.

However, the military writings did not naively assume that solicitude alone would stimulate men to heroic actions against their best
interests—against what the *Tao Te Ching* calls the courage to be daring not to fight—and therefore systematically advanced concepts of
rewards and punishments to stimulate and compel soldiers to the
desired performance. In contrast, the *Tao Te Ching* decries the multi-
plication of laws and the proliferation of punishments, and is thus
surpassed in this regard by the more realistic attitude of the military
writings, including *The Six Secret Teachings*. Compassion and benevolence nurture allegiance and a desire for life, thereby making punishments possible because their deterrent effects depend upon the people fearing death, as observed in chapter 74: "When the people do not
fear death, how will you be able to frighten them with death?" Moreover, as chapter 72 of the *Tao Te Ching* notes, when order disintegrates
and the people become both distressed and depressed, true disaster

will strike: "When the people do not fear awesomeness, then great awesomeness arrives."

Chapter 69 raises an interesting point in this regard:

> No disaster is greater than slighting the enemy, for slighting the enemy approaches the loss of one's treasures. Thus when armies in a standoff attack each other, the one who feels greatest grief [at the conflict] will be victorious.

Solicitude, coupled with the ruler's compassion and natural antipathy at seeing men die, previously mentioned, explains the probable success of armies in such confrontations—not because grief motivates the men to fight, but because the ruler emphasizes their lives and does not want to waste them. This is symptomatic of good government, one that will have acquired the allegiance of the people. However, the military view differs distinctly, tending to decry excessive benevolence in a commander—as the T'ai Kung, among others, observes in "A Discussion of Generals," where one of the ten fatal flaws is "being benevolent but unable to inflict suffering"—and assumes that a certain ruthlessness in the prosecution of battlefield engagements is requisite to campaign success and, paradoxically but unstated, minimizing casualties overall. However, it might be observed that the Chinese view and historical practice was to countenance enormously high casualty rates, with the *Wei Liao-tzu* even stating:

> I have heard that in antiquity those who excelled in employing the army could bear to kill half their officers and soldiers. The next could kill thirty percent, and the lowest ten percent. The awesomeness of one who could sacrifice half of his troops affected all within the Four Seas. The strength of one who could sacrifice thirty percent could be applied to the feudal lords. The orders of one who could sacrifice ten percent would be implemented among his officers and troops.

Unfortunately, a number of World War II engagements illustrated both aspects of this principle.

SIGNS, OMENS, PHASES, AND
PHENOMENAL INTERPRETATION

Concern with perceiving the critical moment and acting appropriately—two distinctly different affairs—may well be said to have been a dominant concern among inhabitants of the late Warring States period. The *I Ching*, developed at least partly in response to the difficulty of interpreting often bizarre crack patterns in tortoise shells, provided a "handbook" method for generating consistent phenomenal signs, although achieving insight into the meaning of these signs remained another matter, prompting the creation of other, more accessible works in the same genre, such as the *Ling Ch'i Ching*. Even though the *I Ching* frequently conceptualizes in terms of military images—the explanations often refer to military events, and it was frequently consulted before battles as early as the Spring and Autumn period, just as the Shang rulers had consulted the turtle oracle—it apparently had no direct influence upon the early military writings. However, the employment of prognosticative phenomena, whether in seeking to harmonize with the universal flux or fathom concrete situations, continued throughout the period and thereafter, despite often strong condemnation by the military writers.

The T'ai Kung's biography relates his effort to stiffen the troops when the omens for attacking the Shang proved inauspicious and heavy wind and rain, both ill portents, arose at the critical moment. To justify his disregard for such overwhelmingly baleful signs, he claimed that the overthrow of the ruling house could hardly produce favorable indications. This may be understood as referring to the moment of action, rather than the enterprise in general, because in "Opening Instructions" he stresses that revolutionary activity cannot be undertaken on personal responsibility alone:

> If there are no ill omens in the Tao of Heaven, you cannot initiate the movement to revolt. If there are no misfortunes in the Tao of Man, your planning cannot precede them. You must first see Heavenly signs, and moreover witness human misfortune; only thereafter can you make plans.

His emphasis upon relieving the suffering of the people, the sole justification in the view of Confucians such as Mencius and the military writers as well, stands forth clearly.

Late in the Warring States period, despite widespread obsession with anticipatory interpretation, the practice of consulting omens was strongly condemned by the *Wei Liao-tzu*, one of the *Seven Military Classics*. Thereafter the military writings witnessed a theoretical bifurcation, with some monographs being devoted to diverse phenomena and their application to the battlefield while such practices were completely eschewed by serious orthodox writings. However, both approaches simultaneously persisted, and in fact the prognosticative writings became intricately fashioned, perceiving not only in actual dust formations useful information about the enemy—such as already appear in *The Art of War*—but in clouds, wind directions, rain patterns, sky colors, and the unusual behavior of animals and birds as well. Two chapters in *The Six Secret Teachings* preserve early materials from this tradition, "The Five Notes" and "The Army's Indications." The former exploits five-phase theory—its only real appearance in the book despite being prominent in *The Art of War*—in terms of the five notes to foretell victory or defeat, while the latter arrays a continuum of signs running from command measures (as evidenced by the drums) to formation characteristics and finally to the behavior of *ch'i* (pneumenal essence) to fathom the enemy. This continuity among various phenomenal manifestations indicates the complex integration of prognosticative practices within the worldview of the military thinkers in the Warring States period.

COPING WITH VICTORY AND SUCCESS

The Taoist vision of Sage administration found throughout the *Tao Te Ching*, sketchily reprised above, emphasizes not only studied inaction, but also actively disdaining, even disclaiming, the fruits of one's efforts. This attitude arises because of a belief in the inherent instability of extremes, translatable as the inescapable decline of positive achievement, particularly when the actor appropriates and dwells in them. Because the strong, the mature, have reached their peak, the Sage essentially hedges against this fated decline by not designating them as such, but allowing for further potential and refusing the static. Naturally this approach is as applicable to military affairs as civil ones, and certainly pertains to battlefield victories, inherently inimical to the Tao even when necessary to preserve the state and its people. Consequently, while not blindly shunning combat, the Taoist commander preserves himself and his victory by distancing himself from the results—for example, through the performance of the rites of mourning, essentially a purification experience that serves to demark the martial from the civil. According to the *Ssu-ma Fa*, an early military text concerned with larger questions as well as tactics and strategy, such ceremonies segregate the two disjunctive spirits of the civil and martial and thereby preserve the innocence of the people while reintegrating the practitioners of violence into the realm of non-violence.

The military thinkers, including the T'ai Kung, although equally troubled by the possibility of becoming enamored by warfare and the fruits of victory, did not believe in essentially disowning the results. Lao-tzu might say, "If one wants to seize the world and act on it, I observe he will not achieve his ends," but *The Six Secret Teachings* advances a more moderate view: "At the extreme, when things return to normal, do not continue to advance and contend, do not withdraw and yield. If you can preserve the state in this fashion, you will share the splendor of Heaven and Earth." However, all the military writings voice concern over prolonged warfare and frequent battles, and while stressing the need to exploit advantage and hard-won victories, also embrace a principle of sufficiency, of stopping when it is appropriate to stop rather than going to extremes. (The difficulty of course lies in distinguishing sufficiency from excess, in not stopping prematurely and inherently thwarting the effort, undermining the effects.) In the T'ai Kung's view, activities, once initiated, need to achieve objectives; otherwise states should refrain from military action. However, even in their pursuit states should behave with humanitarian restraint, and when victorious, preserve the people and the material aspects of civilization. They should fight, as Sun-tzu advised, with the objective of preservation.

Date and Authorship of the Text

The historic T'ai Kung's relationship with *The Six Secret Teachings* remains controversial and is marked by widely differing opinions. The present Chinese title, *T'ai Kung Liu-t'ao,* first appeared in the "Treatise on Literature" incorporated into the *Sui Shu,* the history of the short-lived Sui dynasty written during the T'ang. Prior to this both Liu Pei and the great general Chu-ko Liang are noted as having highly esteemed a book entitled *Liu-t'ao.* Yen Shih-ku, the famous exegete (perhaps erroneously) identified it with another, similarly titled book extant in the Han dynasty, thought to be a Chou dynasty historical writing.

Although the meaning of the title *Liu T'ao* is not completely clear, the first character, *liu,* incontrovertibly means "six." The second character, *t'ao,* has the primary meaning of a "wrap" or "cover"; within a military context it meant the cloth wrapped around a bow, or perhaps a bowcase used to carry it. By extension it means "to conceal" or "to secrete," and by implication it probably came to refer to the skills involved in using a bow in warfare, and therefore military arts in general. Thus the *Liu-t'ao* should be understood as a book containing six categorical discussions about the skills and tactics of warfare. Although it has been virtually ignored in the West, the title has occasionally been translated as the *Six Cases.* However, we have opted to emphasize the aspect of wrapping things, and thereby keeping them secret, together with the putative author's role as a strategist and advisor, and have therefore chosen the title *The Six Secret Teachings.*

Members of the Confucian school, including a number of prominent Sung dynasty scholars, disparaged *The Six Secret Teachings* as a forgery of the Warring States period when the other military writings originated. Thereafter other pedants even attributed it to the T'ang dynasty, vociferously denying it any claim to antiquity whatsoever. Their main criticism focused on the realistic nature of the work and the "despicable policies" which the T'ai Kung clearly advocates. As already mentioned, they dogmatically insisted that true Sages, such as the founders of the Chou dynasty and the T'ai Kung, would not debase themselves nor be compelled to employ artifice, deception, sex, and bribes to achieve their ends. Therefore, from their narrow perspective the conquest of the Shang can only be understood as the victory of culture and Virtue over barbarism and perversity. Unfortunately they systematically ignored the ancient emphasis upon both the civil and martial, and thereby overlooked the decisive nature of the final battle and the conditions preceding it, wherein the vastly outnumbered Chou army, after an extensive forced march, decimated the Shang forces. (A few professional soldiers have contradicted them, emphasizing that the realistic character of the Chou's military activities and their total commitment to employing every means possible to vanquish the evil and preserve the populace should be construed as a clear and certain attestation to the validity of the text.)

Some traditionalists, especially historians with career military service backgrounds apparently anxious to uphold the authenticity of the work, continue to claim it actually dates from the founding of the Chou dynasty. Others of more moderate viewpoint believe that at least some of the historic T'ai Kung's original teachings could have been preserved in terse form on bamboo and orally transmitted by his descendants in the state of Ch'i, providing the foundation for Ch'i's illustrious military studies. However, they grudgingly acknowledge that over the centuries the original discussions probably suffered numerous accretions and losses, admitting that the extant text was probably compiled and revised late in the Warring States period.

Confident assertions that the entire work is a T'ang forgery were dramatically destroyed with the discovery of a virtually identical, although only partial, bamboo slip edition in a Han dynasty tomb in the early 1970s. Coupled with other Han historical references, these remnants prove that portions of the text assumed present form at least by the early Han, and have of course been cited by proponents of the T'ai Kung's inherent connection with the book as evidence for their position. However, even those advocates who staunchly believe a prototype text underlies the present *Six Secret Teachings* are compelled to acknowledge the existence of numerous historical anachronisms. The language and style of writing indicate extensive revisions and perhaps final commitment to written form could not have occurred before the fourth century BCE. The frequent mention of advanced weapons, such as the crossbow and sword, and entire chapters devoted to cavalry tactics, prove the penultimate author lived seven to eight hundred years later than the T'ai Kung. Chapter 55, a remarkable discussion entitled "Equivalent Forces," compares the field effectiveness of the three component forces of chariots, cavalrymen, and infantrymen, although infantry did not become significant until the Spring and Autumn period, and the cavalry only emerged in the third century BCE.

Several scholars have observed that *The Six Secret Teachings* extensively quotes passages and borrows concepts from the other military classics, such as Sun-tzu's famous *Art of War*. However, for most of the classical Chinese writings questions of priority must always be considered a subject for debate. In theory, it might—however unreasonably—be claimed that *The Art of War* is terse and abstract precisely because Sun-tzu benefited from a prior tradition of Chou military thought and, as with the authors of such other works as the *Wei Liao-tzu,* availed himself of concepts from embryonic text of *The Six Secret Teachings* and assiduously assimilated other common sayings. Moreover, in the Warring States period thorough familiarity with all extant military thought would have been a prerequisite for states and commanders to survive. Therefore the absence of both conceptual and textual borrowing would have been even more remarkable than their presence, indicating highly segmented and strictly preserved schools of tactics and secret strategy.

One final viewpoint regarding the text's transmission holds that the famous military writing given to Chang Liang in the turbulent years preceding the Han dynasty's founding was *The Six Secret Teachings* rather than the *Three Strategies of Huang Shih-kung*. This book would be particularly appropriate because of its historical echoes: its readers were committed to the populist overthrow of another brutal, oppressive ruling house, the Ch'in. Accordingly, the suggestion has been made that it was actually composed by a military expert in the third century BCE—no doubt upon common historical information—when the Ch'in was relentlessly destroying its enemies and consolidating its power. This would explain the mature development of concepts and strategies; the extensive knowledge of weapons and defensive equipment; the emphasis upon benevolent government; and the efforts to preserve its secrecy.

I 文韜

CIVIL
SECRET
TEACHING

1 文 師

King Wen's Teacher

KING WEN INTENDED TO GO HUNTING, so Pien, the Scribe, performed divination to inquire about his prospects. The Scribe reported: "While hunting on the north bank of the Wei River you will get a great catch. It will not be any form of dragon, nor a tiger or great bear. According to the signs you will find a duke or marquis there whom Heaven has sent to be your teacher. If employed as your assistant, you will flourish and the benefits will extend to three generations of Chou kings."

King Wen asked: "Do the signs truly signify this?"

The Scribe Pien replied: "My Supreme Ancestor, the Scribe Ch'ou, when performing divination for Sage Emperor Shun, obtained comparable indications. Emperor Shun then found Kao-yao to assist him."

King Wen observed a vegetarian regime for three days to purify himself, then mounted his hunting chariot. Driving his hunting horses, he went out to hunt on the northern bank of the Wei River. Finally he saw the T'ai Kung sitting on a grass mat, fishing. King Wen greeted him courteously and then asked, "Do you take pleasure in fishing?"

The T'ai Kung replied: "The True Man of Worth takes pleasure in attaining his ambitions; the common man takes pleasure in succeeding in his ordinary affairs. Now my fishing is very much like this."

"What do you mean, it is like it?" inquired the king.

The T'ai Kung responded: "In fishing there are three forms of authority: the ranks of salary, death, and offices. Fishing is the means to obtain what you seek. Its nature is deep, and from it much greater principles can be discerned."

King Wen said, "I would like to hear about its nature."

The T'ai Kung elaborated: "When the source is deep, the water flows actively. When the water flows actively, fish spawn there. This is nature. When the roots are deep, the tree is tall. When the tree is tall, fruit is produced. This is nature. When True Men of Worth have sympathies and views in common, they will be drawn together. When they are drawn together, affairs arise. This is nature.

"Speech and response are the adornment of inner emotions. Speaking about true nature is the pinnacle of affairs. Now if I speak about true nature, without avoiding any topic, will you find it abhorrent?"

King Wen replied: "Only a man of true humanity can accept corrections and remonstrance. I have no abhorrence of true nature, so what is your meaning?"

The T'ai Kung said: "When the line is thin and the bait glittering, only small fish will eat it. When the line is heavier and the bait fragrant, medium-sized fish will eat it. But when the line is heavy and the bait generous, large fish will eat it. When the fish take the bait they will be caught on the line. When men take their salary they will submit to the ruler. When you catch fish with bait, the fish can be killed. When you catch men with remuneration, they can be made to exhaust their abilities for you. If you use your family to gain the state, the state can be plucked. If you use your state, the world can be completely acquired.

"Alas, flourishing and florid, although they assemble together they will be scattered! Silent and still, the Sage Ruler's glory will inevitably extend far! Subtle and mysterious, the Virtue of the Sage Ruler as it attracts the people! He alone sees it. Wondrous and joyful, the plans of the Sage Ruler through which everyone seeks and returns to their appropriate places, while he establishes the measures that will gather in their hearts."

King Wen inquired: "How shall we proceed to establish measures so that All under Heaven will give their allegiance?"

The T'ai Kung said: "All under Heaven is not one man's domain. All under Heaven means just that, *all* under Heaven. Anyone who shares profit with all the people under Heaven will gain the world. Anyone who monopolizes its profits will lose the world. Heaven has its seasons, Earth its resources. Being capable of sharing these in common with the people is true humanity. Wherever there is true humanity, All under Heaven will give their allegiance.

"Sparing the people from death; eliminating the hardships of the people; relieving the misfortunes of the people; and sustaining the people in their extremities, is Virtue. Wherever there is Virtue, All under Heaven will give their allegiance.

"Sharing worries, pleasures, likes, and dislikes with the people constitutes righteousness. Where there is righteousness, the people will go.

"In general, people hate death and take pleasure in life. They love Virtue and incline to profit. The ability to produce profit accords with the Tao. Where the Tao resides, All under Heaven will give their allegiance."

King Wen bowed twice and said: "True wisdom! Do I dare not accept Heaven's edict and mandate?"

He had the T'ai Kung ride in the chariot and returned with him, establishing him as his teacher.

2 盈虚

Fullness and Emptiness

KING WEN INQUIRED OF THE T'AI KUNG: "The world is replete with a dazzling array of states—some full, others empty, some well ordered, others in chaos. How does it come to be thus? Is it that the moral qualities of these rulers are not the same? Or that the changes and transformations of the seasons of Heaven naturally cause it to be thus?"

The T'ai Kung said: "If the ruler lacks moral worth, then the state will be in danger and the people in turbulence. If the ruler is a Worthy or Sage, then the state will be at peace and the people well ordered.

Fortune and misfortune lie with the ruler, not with the seasons of Heaven."

King Wen: "May I hear about the Sages of antiquity?"

The T'ai Kung: "Former generations referred to Emperor Yao, in his kingship over the realm in antiquity, as a Worthy ruler."

King Wen: "What was his administration like?"

The T'ai Kung: "When Yao was king of the world he did not adorn himself with gold, silver, pearls, and jade. He did not wear brocaded, embroidered, or elegantly decorated clothes. He did not look at strange, odd, rare, or unusual things. He did not treasure items of amusement nor listen to licentious music. He did not whitewash the walls around the palace or the buildings, nor decoratively carve the beams, square and round rafters, and pillars. He did not even trim the reeds that grew all about his courtyards. He used a deerskin robe to ward off the cold, while simple clothes covered his body. He ate coarse millet and unpolished grains, and thick soups from rough vegetables. He did not, through the untimely imposition of labor service, injure the people's seasons for agriculture and sericulture. He reduced his desires and constrained his will, managing affairs by nonaction.

"He honored the positions of the officials who were loyal, upright, and upheld the laws, and made generous the salaries of those who were pure and scrupulous and loved people. He loved and respected those among the people who were filial and compassionate, and he comforted and encouraged those who exhausted their strength in agriculture and sericulture. Pennants distinguished the virtuous from the evil, being displayed at the gates of the village lanes. He tranquillized his heart and rectified the constraints of social forms. With laws and measures he prohibited evil and artifice.

"Among those he hated, if anyone had merit he would invariably reward him. Among those he loved, if anyone were guilty of an offense he would certainly punish him. He preserved and nurtured the widows, widowers, orphans, and solitary elderly, and gave aid to the families who had suffered misfortune and loss.

"What he allotted to himself was extremely meager, the taxes and services he required of the people extremely few. Thus the myriad peoples were prosperous and happy, and did not have the appearance of suffering from hunger and cold. The hundred surnames revered their ruler as if he were the sun and moon, and gave their emotional allegiance as if he were their father and mother."

King Wen: "Great is the Worthy and Virtuous ruler!"

3

Affairs of State

KING WEN SAID TO THE T'AI KUNG: "I would like to learn about the affair of administering the state. If I want to have the ruler honored and the people settled, how should I proceed?"

The T'ai Kung: "Just love the people."

King Wen: "How does one love the people?"

The T'ai Kung: "Profit them, do not harm them. Help them to succeed, do not defeat them. Give them life, do not slay them. Grant, do not take away. Give them pleasure, do not cause them to suffer. Make them happy, do not cause them to be angry."

King Wen: "May I dare ask you to explain the reasons for these?"

The T'ai Kung: "When the people do not lose their fundamental occupations, you will have profited them. When the farmers do not lose the agricultural seasons, you will have completed them. When you reduce punishments and fines, you give them life. When you impose light taxes, you give to them. When you keep your palaces, mansions, terraces, and pavilions few, you give them pleasure. When the officials are pure and neither irritating nor troublesome, you make them happy.

"But when the people lose their fundamental occupations, you harm them. When the farmers lose the agricultural seasons, you defeat them. When they are innocent but you punish them, you kill them. When you impose heavy taxes, you take from them. When you construct numerous palaces, mansions, terraces, and pavilions, thereby wearing out the people's strength, you make it bitter for them. When the officials are corrupt, irritating, and troublesome, you anger them.

"Thus one who excels at administering a state governs the people as parents govern their beloved children, or as an older brother acts toward his beloved younger brother. When they see their hunger and cold, they are troubled for them. When they see their labors and suffering, they grieve for them.

"Rewards and punishments should be implemented as if being imposed upon yourself. Taxes should be imposed as if taking from yourself. This is the Way to love the people."

4 大禮

The Great Forms of Etiquette

KING WEN ASKED THE T'AI KUNG: "What is the proper form of etiquette between ruler and minister?"

The T'ai Kung said: "The ruler only needs to draw near to the people, subordinates only need to be submissive. He must approach them, not being distant from any. They must be submissive without hiding anything. The ruler wants only to be all-encompassing; subordinates want only to be settled in their positions. If he is all-encompassing he will be like Heaven. If they are settled, they will be like Earth. One Heaven, one Earth—the Great Li is then complete."

King Wen: "How should the ruler act in his position?"

The T'ai Kung: "He should be composed, dignified, and quiet. His softness and self-constraint should be established first. He should excel at giving and not be contentious. He should empty his mind and tranquillize his intentions, awaiting events with uprightness."

King Wen inquired: "How should the ruler listen to affairs?"

The T'ai Kung replied: "He should not carelessly allow them, nor go against opinion and oppose them. If he allows them in this fashion, he will lose his central control; if he opposes them in this way, he will close off his access.

"He should be like the height of a mountain that, when looked up to, cannot be perceived, or the depths of a great abyss that, when measured, cannot be fathomed. Such spiritual and enlightened Virtue is the pinnacle of uprightness and tranquillity."

King Wen inquired: "What should the ruler's wisdom be like?"

The T'ai Kung: "The eye values clarity, the ear values sharpness, the mind values wisdom. If you look with the eyes of All under Heaven, there is nothing you will not see. If you listen with the ears of All under Heaven, there is nothing you will not hear. If you think with the mind of All under Heaven, there is nothing you will not know. When you receive information from all directions just like the spokes converging on the hub of a wheel, your clarity will not be obfuscated."

5 明傳

Clear Instructions

KING WEN, LYING IN BED SERIOUSLY ILL, summoned T'ai Kung Wang and the future King Wu to his side. "Alas, Heaven is about to abandon me. Chou's state altars will soon be entrusted to you. Today I want you, my teacher, to discuss the great principles of the Tao in order to clearly transmit them to my son and grandsons."

The T'ai Kung said: "My king, what do you want to ask about?"

King Wen: "May I hear about the Tao of the former Sages—where it stops, where it begins?"

The T'ai Kung: "If one sees good but is dilatory; if the time for action arrives and one is doubtful; if you know something is wrong but you sanction it—it is in these three that the Tao stops. If one is soft and quiet; dignified and respectful; strong yet genial; tolerant yet hard—it is in these four that the Tao begins. Accordingly, when right-eousness overcomes desire one will flourish; when desire overcomes righteousness one will perish. When respect overcomes dilatoriness it is auspicious; when dilatoriness overcomes respect one is destroyed."

6

Six Preservations

KING WEN ASKED THE T'AI KUNG: "How does the ruler of the state and leader of the people come to lose his position?"

The T'ai Kung said: "He is not cautious about whom he has as asso-ciates. The ruler has 'six preservations' and 'three treasures.'"

King Wen asked: "What are the six preservations?"

The T'ai Kung: "The first is called benevolence, the second right-eousness, the third loyalty, the fourth trust (good faith), the fifth courage, and the sixth planning. These are referred to as the 'six preservations.'"

King Wen asked: "How does one go about carefully selecting men using the six preservations?"

The T'ai Kung: "Make them rich and observe whether they do not commit offenses. Give them rank and observe whether they do not become arrogant. Entrust them with responsibility and see whether they will not change. Employ them and see whether they will not con-ceal anything. Endanger them and see whether they are not afraid. Give them the management of affairs and see whether they are not perplexed.

"If you make them rich but they do not commit offenses, they are benevolent. If you give them rank and they do not grow arrogant, they are righteous. If you entrust them with office and they do not change, they are loyal. If you employ them and they do not conceal anything, they are trustworthy. If you put them in danger and they are not afraid, they are courageous. If you give them the management of affairs and they are not perplexed, they are capable of making plans.

"The ruler must not loan the 'three treasures' to other men. If he loans them to other men, the ruler will lose his awesomeness."

King Wen: "May I ask about the three treasures?"

The T'ai Kung: "Great agriculture, great industry, and great commerce are referred to as the 'three treasures.' If you have the farmers dwell solely in districts of farmers, then the five grains will be sufficient. If you have the artisans dwell solely in districts of artisans, then the implements will be adequate. If you have the merchants dwell solely in districts of merchants, then the material goods will be sufficient.

"If the three treasures are each settled in their places, then the people will not scheme. Do not allow confusion among their districts, do not allow confusion among their clans. Ministers should not be more wealthy than the ruler. No other cities should be larger than the ruler's state capital. When the six preservations are fully implemented, the ruler will flourish. When the three treasures are complete, the state will be secure."

7

Preserving the State's Territory

KING WEN ASKED THE T'AI KUNG: "How does one preserve the state's territory?"

The T'ai Kung: "Do not estrange your relatives. Do not neglect the masses. Be conciliatory and solicitous toward nearby states, and control the four quarters.

"Do not loan the handles of state to other men. If you loan the handles of state to other men then you will lose your authority. Do not dig valleys deeper to increase hills. Do not abandon the foundation to govern the branches. When the sun is at midday, you should dry things. If you grasp a knife you must cut. If you hold an axe you must attack.

"If, at the height of day, you do not dry things in the sun, this is termed losing the time. If you grasp a knife but do not cut anything, you will lose the moment for profits. If you hold an axe but do not attack, then bandits will come.

"If trickling streams are not blocked they will become great rivers. If you do not extinguish the smallest flames, what will you do about a great conflagration? If you do not eliminate the two-leaf sapling, how will you use your ax when the tree has grown?

"For this reason the ruler must focus upon developing wealth within his state. Without material wealth he has nothing with which to be benevolent. If he does not bespread beneficence he will have nothing with which to bring his relatives together. If he estranges his relatives it will be harmful. If he loses the common people he will be defeated.

"Do not loan sharp weapons to other men. If you loan sharp weapons to other men, you will be hurt by them, and will not live out your allotted span of years."

King Wen said: "What do you mean by benevolence and righteousness?"

The T'ai Kung: "Respect the common people, unite your relatives. If you respect the common people they will be at peace. And if you unite your relatives they will be happy. This is the way to implement the essential cords of benevolence and righteousness.

"Do not allow other men to snatch away your awesomeness. Rely on your wisdom, follow the constant. Those that submit and accord with you, treat generously with Virtue. Those that oppose you, break with force. If you respect the people and are decisive, then All under Heaven will be peaceful and submissive."

8  Preserving the State

KING WEN ASKED THE T'AI KUNG: "How does one preserve the state?"

The T'ai Kung: "You should observe a vegetarian fast because I am about to speak to you about the essential principles of Heaven and Earth, what the four seasons produce, the Tao of true humanity and sagacity, and the nature of the people's impulses."

The king observed a vegetarian regime for seven days, then, facing north, bowed twice and requested instruction.

The T'ai Kung said: "Heaven gives birth to the four seasons, Earth produces the myriad things. Under Heaven there are the people, and the Sage acts as their shepherd.

"Thus the Tao of spring is birth and the myriad things begin to flourish. The Tao of summer is growth, the myriad things mature. The Tao of autumn is gathering, the myriad things are full. The Tao of winter is storing away, the myriad things are still. When they are full they are stored away; after they are stored away they again revive. No one knows where it ends, no one knows where it begins. The Sage accords with it, and models himself on Heaven and Earth. Thus when the realm is well ordered, his benevolence and sagacity are hidden. When All under Heaven are in turbulence, his benevolence and sagacity flourish. This is the true Tao.

"In his position between Heaven and Earth, what the Sage treasures is substantial and vast. Relying on the constant to view it, the people are at peace. But when the people are agitated it creates impulses. When impulses stir, conflict over gain and loss arises. Thus it is initiated in *yin*, but coalesces in *yang*. If someone ventures to be the first leader, All under Heaven will unite with him. At the extreme, when things return to normal, do not continue to advance and contend, do not withdraw and yield. If you can preserve the state in this fashion, you will share the splendor of Heaven and Earth."

9 上賢

Honoring the Worthy

KING WEN ASKED THE T'AI KUNG: "Among those I rule, who should be elevated, who should be placed in inferior positions? Who should be selected for employment, who cast aside? How should they be restricted, how stopped?"

The T'ai Kung said: "Elevate the Worthy, and place the unworthy in inferior positions. Choose the sincere and trustworthy, eliminate the deceptive and artful. Prohibit violence and turbulence, stop extravagance and ease. Accordingly, one who exercises kingship over the people recognizes 'six thieves' and 'seven harms.'"

King Wen said: "I would like to know about its Tao."

The T'ai Kung: "As for the 'six thieves':

"First, if your subordinates build large palaces and mansions, pools and terraces, and amble about enjoying the pleasures of scenery and female musicians, it will injure the king's Virtue.

"Second, when the people are not engaged in agriculture and sericulture, but instead give rein to their tempers and travel about as bravados, disdaining and transgressing the laws and prohibitions, not following the instructions of the officials, it harms the king's transforming influence.

"Third, when officials form cliques and parties, obfuscating the worthy and wise, obstructing the ruler's clarity, it injures the king's authority.

"Fourth, when the knights are contrary-minded and conspicuously display 'high moral standards'—taking such behavior to be powerful expression of their *ch'i*, and have private relationships with other feudal lords, slighting their own ruler, it injures the king's awesomeness.

"Fifth, when subordinates disdain titles and positions, are contemptuous of the administrators, and are ashamed to face hardship for their ruler, it injures the efforts of the meritorious subordinates.

"Sixth, when strong clans encroach upon others, seizing what they want, insulting and ridiculing the poor and weak, it injures the work of the common people.

"The 'seven harms':

"First, men without knowledge or strategic planning ability are generously rewarded and honored with rank. Therefore the strong and courageous who regard war lightly take their chances in the field. The king must be careful not to employ them as generals.

"Second, they have reputation but lack substance. What they say is constantly shifting. They conceal the good and point out deficiencies. They view advancement and dismissal as a question of skill. The king should be careful not to make plans with them.

"Third, they make their appearance simple, wear ugly clothes, speak about actionless action in order to seek fame, and talk about nondesire in order to gain profit. They are artificial men, and the king should be careful not to bring them near.

"Fourth, they wear strange caps and belts, and their clothes are overflowing. They listen widely to the disputations of others and speak speciously about unrealistic ideas, displaying them as a sort of personal adornment. They dwell in poverty and live in tranquillity, deprecating the customs of the world. They are cunning people, and the king should be careful not to favor them.

"Fifth, with slander, obsequiousness, and pandering they seek office and rank. They are courageous and daring, treating death lightly, out of their greed for salary and position. They are not concerned with major affairs, but move solely out of avarice. With lofty talk and specious discussions they please the ruler. The king should be careful not to employ them.

"Sixth, they have buildings elaborately carved and inlaid. They promote artifice and flowery adornment to the injury of agriculture. You must prohibit them.

"Seventh, they create magical formulas and weird techniques, practice sorcery and witchcraft, advance unorthodox ways, and circulate inauspicious sayings, confusing and befuddling the good people. The king must stop them.

"Now when the people do not exhaust their strength, they are not our people. If the officers are not sincere and trustworthy, they are not our officers. If the ministers do not offer loyal remonstrance, they are not our ministers. If the officials are not even-handed, pure, and love the people, they are not our officials. If the chancellor cannot enrich the state and strengthen the army, harmonize *yin* and *yang*, and ensure security for the ruler of a state of ten thousand chariots—and moreover properly control the ministers, set names and realities, make clear rewards and punishments, and give pleasure to the people—he is not our chancellor.

"Now the Tao of the king is like that of a dragon's head. He dwells in the heights and looks out far. He sees deeply and listens carefully. He displays his form, but conceals his nature. He is like the heights of Heaven, which cannot be perceived. He is like the depths of an abyss that cannot be fathomed. Thus if he should get angry but does not, evil subordinates will arise. If he should execute but does not, great thieves will appear. If strategic military power is not exercised, enemy states will grow strong."

King Wen said: "Excellent!"

10 舉 賢

Advancing the Worthy

KING WEN ASKED THE T'AI KUNG: "How does it happen that a ruler may exert himself to advance the Worthy, but is unable to obtain any results from such efforts, and in fact the world grows increasingly turbulent, even to the point that he is endangered or perishes?"

The T'ai Kung: "If one advances the Worthy but does not employ them, this is attaining the name of 'advancing the Worthy' but lacking the substance of 'using the Worthy.'"

King Wen asked: "Whence comes the error?"

The T'ai Kung: "The error lies in wanting to employ men who are popularly praised rather than obtaining true Worthies."

King Wen: "How is that?"

The T'ai Kung said: "If the ruler takes those that the world commonly praises as being Worthies, and those that it condemns as being worthless, then the larger cliques will advance and the smaller ones will retreat. In this situation groups of evil individuals will associate together to obscure the Worthy. Loyal subordinates will die even though innocent. And perverse subordinates will obtain rank and position through empty fame. In this way, as turbulence continues to grow in the world, the state cannot avoid danger and destruction."

King Wen asked: "How does one advance the Worthy?"

The T'ai Kung replied: "Your general and chancellor should divide the responsibility, each of them selecting men based upon the names of the positions. In accord with the name of the position, they will assess the substance required. In selecting men, they will evaluate their abilities, making the reality of their talents match the name of the position. When the name matches the reality, you will have realized the Tao for advancing the Worthy."

11

Rewards and Punishments

KING WEN ASKED THE T'AI KUNG: "Rewards are the means to preserve the encouragement of good, punishments the means to display the rectification of evil. By rewarding one man I want to stimulate a hundred, by punishing one man rectify the multitude. How can I do it?"

The T'ai Kung said: "In general, in employing rewards one values credibility; in employing punishments one values certainty. When rewards are trusted and punishments inevitable wherever the eye sees and the ear hears, then even where they do not see or hear there is no one who will not be transformed in their secrecy. Since the ruler's sincerity extends to Heaven and Earth, and penetrates to the spirits, how much the more so to men?"

12

The Tao of the Military

KING WU ASKED THE T'AI KUNG: "What is the Tao of the military?"

The T'ai Kung said: "In general, as for the Tao of the military, nothing surpasses unity. The unified can come alone, can depart alone. The Yellow Emperor said, 'Unification approaches the Tao and touches upon the spiritual.' Its employment lies in the subtle; its conspicuous manifestation lies in the strategic configuration of power; its completion lies with the ruler. Thus the Sage Kings termed weapons evil implements, but when they had no alternative, they employed them.

"Today the Shang King knows about existence, but not about perishing. He knows pleasure, but not disaster. Now existence does not lie in existence, but in thinking about perishing. Pleasure does not lie in pleasure, but in contemplating disaster. Now that you have already pondered the source of such changes, why do you trouble yourself about the future flow of events?"

King Wu said: "Suppose two armies encounter each other. The enemy cannot come forward, and we cannot go forward. Each side goes about establishing fortifications and defenses, without daring to be the first to attack. If I want to launch a sudden attack, but lack any tactical advantage, what should I do?"

The T'ai Kung said: "Make an outward display of confusion while actually being well ordered. Show an appearance of hunger while actually being well fed. Keep your sharp weapons within, and show only dull and poor weapons outside. Have some troops come together, others split up; some assemble, others scatter. Make secret plans, keep your intentions secret. Raise the height of fortifications, and conceal your elite troops. If the officers are silent, not making any sounds, the enemy will not know our preparations. Then if you want to take his western flank, attack the eastern one."

King Wu said: "If the enemy knows my true situation and has penetrated my plans, what should I do?"

The T'ai Kung said: "The technique for military conquest is to carefully investigate the enemy's intentions and quickly take advantage of them, launching a sudden attack where unexpected.

II 武韜

MARTIAL
SECRET
TEACHING

13 發 啓

Opening Instructions

KING WEN, IN THE CAPITAL OF FENG, summoned the T'ai Kung. "Alas! The Shang King is extremely perverse, judging the innocent guilty and having them executed. If you assist me in my concern for these people, how might we proceed?"

The T'ai Kung replied: "You should cultivate your Virtue, submit to the guidance of Worthy men, extend beneficence to the people, and observe the Tao of Heaven. If there are no ill omens in the Tao of Heaven, you cannot initiate the movement to revolt. If there are no misfortunes in the Tao of Man, your planning cannot precede them. You must first see Heavenly signs, and moreover witness human misfortune; only thereafter can you make plans. You must look at the Shang King's *yang* aspects, and moreover his *yin* side, for only then will you know his mind. You must look at his external activities, and moreover his internal ones, for only then will you know his thoughts. You must observe those distant from him, and also observe those close to him, for only then will you know his emotions.

"If you implement the Tao, the Tao can be attained. If you enter by the gate, the gate can be entered. If you set up the proper forms of etiquette, the forms can be perfected. If you fight with the strong, the strong can be conquered. If you can attain complete victory without fighting, without the great army suffering any losses, you will have penetrated even the realm of ghosts and spirits. How marvelous! How subtle!

"If you suffer the same illness as other people, and you all aid each other; if you have the same emotions and complete each other, the same hatreds and assist each other, and the same likes and seek them together—then without any armored soldiers you will win, without any battering rams you will have attacked, and without moats and ditches you will have defended.

"The greatest wisdom is not wise; the greatest plans not planned; the greatest courage not courageous; the greatest gain not profitable. If you profit All under Heaven, All under Heaven will be open to you. If you harm All under Heaven, All under Heaven will be closed. All under Heaven is not the property of one man, but of All under Heaven. If you take All under Heaven as if pursuing some wild animal, then All under Heaven will want to carve up the realm like a piece of meat. If you all ride in the same boat to cross over the water, after completing the crossing you will all have profited. However, if you fail to make the crossing, then you will all suffer the harm. If you act as if you are all on the same vessel the empire will be open to your aim, and none will be closed to you.

"He who does not take from the people, takes the people. He who does not take from the people, the people will profit. He who does not take from the states, the states will profit. He who does not take from All under Heaven, All under Heaven will profit. Thus the Tao lies in what cannot be seen, affairs lie in what cannot be heard, and victory lies in what cannot be known. How marvelous! How subtle!

"When an eagle is about to attack, it will fly low and draw in its wings. When a fierce wild cat is about to strike, it will lay back its ears and crouch down low. When the Sage is about to move, he will certainly display a stupid countenance.

"Now there is the case of Shang, where the people muddle and confuse each other. Mixed up and extravagant, their love of pleasure and sex is endless. This is a sign of a doomed state. I have observed their fields—weeds and grass overwhelm the crops. I have observed their people—the perverse and crooked overcome the straight and upright. I have observed their officials—they are violent, perverse, inhumane, and evil. They overthrow the laws and make a chaos of the punishments. Neither the upper nor lower ranks have awakened to this state of affairs. It is time for their state to perish.

"When the sun appears the myriad things are all illuminated. When great righteousness appears the myriad things all profit. When the great army appears the myriad things all submit. Great is the Virtue of the Sage! Listening by himself, seeing by himself, this is his pleasure!"

14 文 啓

Civil Instructions

KING WEN ASKED THE T'AI KUNG: "What does the Sage preserve?"

The T'ai Kung said: "What worries does he have? What constraints? The myriad things all naturally realize their positions. What constraints, what worries? The myriad things all flourish. No one realizes the transforming influence of government; moreover, no one realizes the effects of the passing of time. The Sage preserves the Tao of actionless action and the myriad things are transformed. What is exhausted? When things reach the end they return again to the beginning. Relaxed and complacent the Sage turns about, seeking it. Seeking it he gains it, and cannot but store it. Having already stored it, he cannot but implement it. Having already implemented it, he does not turn about and make it clear that he did so. Now, because Heaven and Earth do not illuminate themselves they are forever able to give birth to the myriad things. The Sage does not cast light upon himself so he is able to attain a glorious name.

"The Sages of antiquity assembled people to comprise families; assembled families to compose states; and assembled states to constitute the realm of All under Heaven. They divided the realm and enfeoffed Worthy men to administer the states. They officially designated this order as the Great Outline.

"They promulgated the government's instructions and accorded with the people's customs. They transformed the multitude of crooked into the straight, changing their form and appearance. Although the customs of the various states were not the same, they all took pleasure in their respective places. The people loved their rulers, so it was termed the Great Settlement.

"Ah, the Sage concentrates upon tranquillizing them, the Worthy focuses upon rectifying them. The stupid man cannot be upright; therefore, he contends with other men. When the ruler labors, punishments become numerous. When punishments are numerous, the people are troubled. When the people are troubled, they leave and wander off. No one, irrespective of his position, can be settled in his life, and generations on end have no rest. This they termed the Great Loss.

"The people of the world are like flowing water. If you obstruct it, it will stop. If you open a way, it will flow. If you keep it quiet, it will be clear. How spiritual! When the Sage sees the beginning, he knows the end."

King Wen said: "How does one tranquillize them ?"

The T'ai Kung: "Heaven has its constant forms, the people have their normal lives. If you share life with All under Heaven, then All under Heaven will be tranquil. The pinnacle accords with them, the next highest transforms them. When the people are transformed and follow their government, then Heaven takes no action but affairs are complete. The people do not give anything to the ruler, so are enriched of themselves. This is the Virtue of the Sage."

King Wen: "What my lord has said accords with what I embrace. From dawn to night I will think about it, never forgetting it, employing it as our constant principle."

15 文 伐

Civil Offensive

KING WEN ASKED THE T'AI KUNG: "What are the methods for civil offensives?"

The T'ai Kung replied: "There are twelve measures for civil offensives.

"First, accord with what he likes in order to accommodate his wishes. He will eventually grow arrogant, and invariably mount some perverse affair. If you can appear to follow along, you will certainly be able to eliminate him.

"Second, become familiar with those he loves in order to fragment his awesomeness. When men have two different inclinations, their loyalty invariably declines. When his court no longer has any loyal ministers, the state altars will inevitably be endangered.

"Third, covertly bribe his assistants, fostering a deep relationship with them. While they will bodily stand in his court, their emotions will be directed outside it. The state will certainly suffer harm.

"Fourth, assist him in his licentiousness and indulgence in music in order to dissipate his will. Make him generous gifts of pearls and jade, and ply him with beautiful women. Speak deferentially, listen respectfully, follow his commands, and accord with him in everything. He will never imagine you might be in conflict with him. Our treacherous measures will then be settled.

"Fifth, treat his loyal officials very generously, but reduce the gifts you provide to the ruler. Delay his emissaries, do not listen to their missions. When he eventually dispatches other men treat them with sincerity, embrace and trust them. The ruler will then again feel you are in harmony with him. If you manage to treat his formerly loyal officials very generously, his state can then be plotted against.

"Sixth, make secret alliances with his favored ministers, but visibly keep his less-favored outside officials at a distance. His talented people will then be under external influence, while enemy states encroach upon his territory. Few states in such a situation have survived.

"Seventh, if you want to bind his heart to you, you must offer generous presents. To gather in his assistants, loyal associates, and loved ones, you must secretly show them the gains they can realize by colluding with you. Have them slight their work, and then their preparations will be futile.

"Eighth, gift him with great treasures and make plans with him. When the plans are successful and profit him, he will have faith in you because of the profits. This is what is termed 'being closely embraced.' The result of being closely embraced is that he will inevitably be used by us. When someone rules a state but is externally controlled, his territory will inevitably be defeated.

"Ninth, honor him with praise. Do nothing that will cause him personal discomfort. Display the proper respect accruing to a great power, and your obedience will certainly be trusted. Magnify his honor, being the first to gloriously praise him, humbly embellishing him as a Sage. Then his state will suffer great loss!

"Tenth, be submissive so that he will trust you, and thereby learn about his true situation. Accept his ideas and respond to his affairs as if you were twins. Once you have learned everything, subtly gather in his power. Thus when the ultimate day arrives it will seem as if Heaven itself destroyed him.

"Eleventh, block up his access by means of the Tao. Among subordinates there is no one who does not value rank and wealth, nor hate danger and misfortune. Secretly express great respect toward them, and gradually bestow valuable gifts in order to gather in the more outstanding talents. Accumulate your own resources until they become very substantial, but manifest an external appearance of shortage. Covertly bring in wise knights, and entrust them with planning great strategy. Attract courageous knights and augment their spirit. Even when they are sufficiently rich and honored continue to increase them. When your faction has been fully established you will have attained the objective referred to as 'blocking his access.' If someone has a state, but his access is blocked, how can he be considered as having the state?

"Twelfth, support his dissolute officials in order to confuse him. Introduce beautiful women and licentious sounds in order to befuddle him. Send him outstanding dogs and horses in order to tire him. From time to time allow him great power in order to entice him to greater arrogance. Then investigate Heaven's signs and plot with the world against him.

"When these twelve measures are fully employed they will become a military weapon. Thus when, as it is said, one 'looks at Heaven above and investigates Earth below' and the proper signs are already visible, attack him."

16 順啓

Instructions on According with the People

KING WEN ASKED THE T'AI KUNG: "What should one do so that he can govern All under Heaven?"

The T'ai Kung said: "When your greatness overspreads All under Heaven, only then will you be able to encompass it. When your trustworthiness has overspread All under Heaven, only then will you be able to make covenants with it. When your benevolence has overspread All under Heaven, only then will you be able to embrace it. When your grace has overspread All under Heaven, only then can you preserve it. When your authority covers the world, only then will you be able not to lose it. If you govern without doubt, then the revolutions of Heaven will not be able to shift your rule, nor the changes of the seasons be able to affect it. Only when these six are complete will you be able to establish a government for All under Heaven.

"Accordingly, one who profits All under Heaven will find All under Heaven open to him. One who harms All under Heaven will find All under Heaven closed to him. If one gives life to All under Heaven, All under Heaven will regard him as Virtuous. If one kills All under Heaven, All under Heaven will regard him as a brigand. If one penetrates to All under Heaven, All under Heaven will be accessible to him; if one impoverishes All under Heaven, All under Heaven will regard him as their enemy. One who gives peace to All under Heaven, All under Heaven will rely on; one who endangers All under Heaven, All under Heaven will view as a disaster. All under Heaven is not the realm of one man. Only one who possesses the Tao can dwell in the position of authority."

17

Three Doubts

KING WU INQUIRED OF THE T'AI KUNG: "I want to attain our aim of overthrowing the Shang, but I have three doubts. I am afraid that our strength will be inadequate to attack the strong, estrange his close supporters within the court, and disperse his people. What should I do?"

The T'ai Kung replied: "Accord with the situation, be very cautious in making plans, and employ your material resources. Now in order to attack the strong you must nurture them to make them even stronger, and increase them to make them even more extensive. What is too strong will certainly break; what is too extended must have deficiencies. Attack the strong through his strength. Cause the estrangement of his favored officials by using his favorites, and disperse his people by means of the people.

"Now in the Tao of planning, thoroughness and secrecy are treasured. You should become involved with him in numerous affairs, and ply him with temptations of profit. Conflict will then surely arise.

"If you want to cause his close supporters to become estranged from him, you must do it by using what they love, making gifts to those he favors, giving them what they want. Tempt them with what they find profitable, thereby making them ambitious. Those who covet profits will be extremely happy at the prospects, and their remaining doubts will be ended.

"Now without doubt, the Tao for attacking is to first obfuscate the king's clarity, and then to attack his strength, destroying his greatness and eliminating the misfortune of the people. Debauch him with beautiful women, entice him with profit. Nurture him with flavors, and provide him with the company of female musicians. Then after you have caused his subordinates to become estranged from him, you must cause the people to grow distant from him, while never letting him know your plans. Appear to support him and draw him into your trap. Do not let him become aware of what is happening, for only then can your plan be successful.

"When bestowing your beneficence upon the people you cannot begrudge the expense. The people are like cows and horses. Frequently make gifts of food and clothing, and follow up by loving them.

"The mind is the means to open up knowledge; knowledge the means to open up the source of wealth; and wealth the means to open up the people. Gaining the allegiance of the people is the way to attract Worthy men. When one is enlightened by Sagely advisors he can become king of all the world."

III 龍韜

DRAGON
SECRET
TEACHING

18 王翼

The King's Wings

KING WU ASKED THE T'AI KUNG: "When the king commands the army he must have 'legs and arms' and 'feathers and wings' to bring about his awesomeness and spiritualness. How should this be done?"

The T'ai Kung said: "Whenever one mobilizes the army it takes the commanding general as its fate. Its fate lies in penetrating understanding of all aspects, not in clinging to one technique. In accord with their abilities assign duties, each one taking charge of what they are good at, constantly changing and transforming with the times, to create the essential principles and order. Thus the general has seventy-two legs and arms and feathers and wings in order to respond to the Tao of Heaven. Prepare their number according to method, being careful that they know its orders and principles. When you have all the different abilities and various skills, then the myriad affairs will be complete."

King Wu asked: "May I ask about the various categories?"

The T'ai Kung said: "Chief of Planning, one: in charge of advising about secret plans for responding to sudden events; investigating Heaven so as to eliminate sudden change; exercising general supervision over all planning; and protecting and preserving the lives of the people.

"Planning Officers, five: responsible for planning security and danger; anticipating the unforeseen; discussing performance and ability; making clear rewards and punishments; appointing officers; deciding the doubtful; and determining what is advisable and what is not.

"Astrologers, three: responsible for the stars and calendar; observing the wind and *ch'i*; predicting auspicious days and times; investigating signs and phenomena; verifying disasters and abnormalities; and knowing Heaven's mind with regard to the moment for completion or abandonment.

"Topographers, three: in charge of the army's disposition and strategic configuration of power when moving and stopped; information on strategic advantages and disadvantages; precipitous and easy passages, both near and far; and water and dry land, mountains and defiles, so as not to lose the advantages of terrain.

"Strategists, nine: responsible for discussing divergent views; analyzing the probable success or failure of various operations; selecting the weapons and training men in their use; and identifying those who violate the ordinances.

"Supply Officers, four: responsible for calculating the requirements for food and water; preparing the foodstocks and supplies; transporting provisions along the route; and supplying the five grains so as to ensure that the army will not suffer any hardship or shortage.

"Officers for Flourishing Awesomeness, four: responsible for picking men of talent and strength; for discussing weapons and armor; for setting up attacks that race like the wind and strike like thunder, so that the enemy does not know where they come from.

"Secret Signals Officers, three: responsible for the pennants and drums, for clearly signaling to the eyes and ears; for creating deceptive signs and seals; issuing false designations and orders; and for stealthily and hastily moving back and forth, going in and out like spirits.

"Legs and Arms, four: responsible for undertaking heavy duties and handling difficult tasks; for the repair and maintenance of ditches and moats; and for keeping the walls and ramparts in repair, in order to defend against and repel the enemy.

"Liaison Officers, two: responsible for gathering what has been lost and supplementing what is in error; for receiving honored guests; holding discussions and talks; mitigating disasters; and resolving difficulties.

"Officers of Authority, three: responsible for implementing the unorthodox and deceptive; for establishing the different and the unusual, things that people do not recognize; and for putting into effect inexhaustible transformations.

"Ears and Eyes, seven: responsible for going about everywhere, listening to what people are saying; seeing the changes; and observing the officers in all four directions, and the army's true situation.

"Claws and Teeth, five: responsible for raising awesomeness and martial spirit; for stimulating and encouraging the Three Armies, causing them to risk hardship and attack the enemy's elite troops without ever having any doubts or second thoughts.

"Feathers and Wings, four: responsible for flourishing the name and fame of the army; for shaking distant lands with its image; and moving all within the four borders in order to weaken the enemy's spirit.

"Roving Officers, eight: responsible for spying upon the enemy's licentiousness and observing their changes; manipulating their emotions; and observing the enemy's thoughts in order to act as spies.

"Officers of Techniques, two: responsible for spreading slander and falsehoods, and for calling upon ghosts and spirits, in order to confuse the minds of the populace.

"Officers of Prescriptions, three: in charge of the hundred medicines; managing blade wounds; and curing the various maladies.

"Accountants, two: responsible for accounting for the provisions and foodstuffs within the Three Armies' encampments and ramparts; for the fiscal materials employed; and for receipts and disbursements."

19 論 將

A Discussion of Generals

KING WU ASKED THE T'AI KUNG: "What should a general be?"

The T'ai Kung replied: "Generals have five critical talents and ten excesses."

King Wu said: "Dare I ask you to enumerate them?"

The T'ai Kung elaborated: "What we refer to as the five talents are courage, wisdom, benevolence, trustworthiness, and loyalty. If he is courageous he cannot be overwhelmed. If he is wise he cannot be forced into turmoil. If he is benevolent he will love his men. If he is trustworthy he will not be deceitful. If he is loyal he will not be of two minds.

"What are referred to as the ten errors are as follows: being courageous and treating death lightly; being hasty and impatient; being greedy and loving profit; being benevolent but unable to inflict suffering; being wise but afraid; being trustworthy and liking to trust others; being scrupulous and incorruptible but not loving men; being wise but indecisive; being resolute and self-reliant; and being fearful while liking to entrust responsibility to other men.

"One who is courageous and treats death lightly can be destroyed by violence. One who is hasty and impatient can be destroyed by persistence. One who is greedy and loves profit can be bribed. One who is benevolent but unable to inflict suffering can be worn down. One who is wise but fearful can be distressed. One who is trustworthy and likes to trust others can be deceived. One who is scrupulous and incorruptible but does not love men can be insulted. One who is wise but indecisive can be suddenly attacked. One who is resolute and self-reliant can be confounded by events. One who is fearful and likes to entrust responsibility to others can be tricked.

"Thus, warfare is the greatest affair of state, the Tao of survival or extinction. The fate of the state lies in the hands of the general. The general is the support of the state, a man that the former kings all valued. Thus in commissioning a general you cannot but carefully evaluate and investigate his character.

"Thus it is said that two armies will not be victorious, nor will both be defeated. When the army ventures out beyond the borders, before they have been out ten days, even if a state has not perished, one army will certainly have been destroyed and the general killed."

King Wu: "Marvelous!"

20 選 将

Selecting Generals

KING WU ASKED THE T'AI KUNG: "If a king wants to raise an army, how should he go about selecting and training heroic officers, and determining their moral qualifications?"

The T'ai Kung said: "There are fifteen cases where a knight's external appearance and internal character do not cohere. These are:

"He appears to be a Worthy, but is immoral.

"He seems warm and conscientious, but is a thief.

"His countenance is reverent and respectful, but his heart is insolent.

"Externally he is incorruptible and circumspect, but he lacks respect.

"He appears perceptive and sharp, but lacks such talent.

"He appears profound, but lacks all sincerity. .

"He appears adept at planning, but is indecisive.

"He appears decisive and daring, but is incapable.

"He appears guileless, but is not trustworthy.

"He appears confused and disoriented, but on the contrary is loyal and substantial.

"He appears to engage in specious discourse, but is a man of merit and achievement.

"He appears courageous, but is afraid.

"He seems severe and remote, but on the contrary easily befriends men.

"He appears forbidding, but on the contrary is quiet and sincere.

"He appears weak and insubstantial, yet when dispatched outside the state there is nothing he does not accomplish, no mission that he does not execute successfully.

"Those whom the world disdains the Sage values. Ordinary men do not know these things, only great wisdom can discern the edge of these matters. This is because the knight's external appearance and internal character do not visibly cohere."

King Wu asked: "How does one know this?"

The T'ai Kung replied: "There are eight forms of evidence by which you may know it.

"First, question them and observe the details of their reply.

"Second, verbally confound and perplex them and observe how they change.

"Third, discuss things that you have secretly learned to observe their sincerity.

"Fourth, clearly and explicitly question them to observe their virtue.

"Fifth, appoint them to positions of financial responsibility to observe their honesty.

"Sixth, test them with beautiful women to observe their uprightness.

"Seventh, confront them with difficulties to observe their courage.

"Eighth, get them drunk to observe their deportment.

"When all eight have been fully explored, then the Worthy and unworthy can be distinguished."

21 立 将

Appointing the General

KING WU ASKED THE T'AI KUNG: "What is the Tao for appointing the commanding general?"

The T'ai Kung said: "When the state encounters danger, the ruler should vacate the Main Hall, summon the general, and charge him as follows: 'The security or endangerment of the Altars of State all lie with the army's commanding general. At present such-and-such a state does not act properly submissive. I would like you to lead the army forth to respond to it.'

"After the general has received his mandate, command the Grand Scribe to bore the sacred tortoiseshell to divine an auspicious day. Thereafter, to prepare for the chosen day, observe a vegetarian regime for three days, and then go to the ancestral temple to hand over the *fu* and *yüeh* axes.

"After the ruler has entered the gate to the temple, he stands facing west. The general enters the temple gate and stands facing north. The ruler personally takes the *yüeh* axe, and holding it by the head, passes the handle to the general, saying, 'From this to Heaven above will be controlled by the General of the Army.' Then taking the *fu* axe by the handle he should give the blade to the general, saying, 'From this to the depths below will be controlled by the General of the Army. When you see a vacuity in the enemy you should advance; when you see substantiality you should halt. Do not assume that the Three Armies are large and treat the enemy lightly. Do not commit yourself to die just because you have received a heavy responsibility. Do not, because you are honored, regard other men as lowly. Do not rely upon yourself alone and contravene the masses. Do not take verbal facility to be a sign of certainty. When the officers have not yet been seated, do not sit. When the officers have not yet eaten, do not eat. You should share heat and cold with them. If you behave in this way the officers and masses will certainly exhaust their strength in fighting to the death.'

"After the general has received his mandate, he bows and responds to the ruler: 'I have heard that a country cannot follow the commands of another state's government, while an army in the field cannot follow central government control. Someone of two minds cannot properly serve his ruler, someone in doubt cannot respond to the enemy. I have already received my mandate, and taken sole control of the awesome power of the *fu* and *yüeh* axes. I dare not return alive. I would like to request that you condescend to grant complete and sole command to me. If you do not permit it, I dare not accept the post of general.' The king then grants it, and the general formally takes his leave and departs.

"Military matters are not determined by the ruler's commands; they all proceed from the commanding general. When the commanding general approaches an enemy and decides to engage in battle, he is not of two minds. In this way there is no Heaven above, no Earth below, no enemy in front, and no ruler to the rear. For this reason the wise make plans for him, the courageous fight for him. Their spirit soars to the blue clouds, they are swift like galloping steeds. Even before the blades clash, the enemy surrenders submissively.

"War is won outside the borders of the state, but the general's merit is established within it. Officials are promoted and receive the highest rewards; the hundred surnames rejoice; and the general is blameless. For this reason the winds and rains will be seasonable; the five grains will grow abundantly; and the altars of state will be secure and peaceful."

King Wu said: "Excellent."

22 将威

The General's Awesomeness

KING WU ASKED: "How does the general create awesomeness? How can he be enlightened? How can he make his prohibitions effective and get his orders implemented?"

The T'ai Kung said: "The general creates awesomeness by executing the great, and becomes enlightened by rewarding the small. Prohibitions are made effective and laws implemented by careful scrutiny in the use of punishments. Therefore, if by executing one man the entire army will quake, kill him. If by rewarding one man the masses will be pleased, reward him.

"In executing, value the great; in rewarding, value the small. When you kill the powerful and the honored, this is punishment that reaches the pinnacle. When rewards extend down to the cowherds, grooms, and stablemen, these are rewards penetrating downward to the lowest. When punishments reach the pinnacle and rewards penetrate to the lowest, then your awesomeness has been effected."

23 勵軍

Encouraging the Army

KING WU ASKED THE T'AI KUNG: "When we attack I want the masses of the Three Armies to contend with each other to scale the wall first, and compete with each other to be in the forefront when we fight in the field. When they hear the sound of the gongs to retreat they will be angry, and when they hear the sound of the drums to advance they will be happy. How can we accomplish this?"

The T'ai Kung said: "A general has three techniques for attaining victory."

King Wu asked: "May I ask what they are?"

The T'ai Kung: "If in winter the general does not wear a fur robe, in summer he does not carry a fan, and in the rain he does not set up a canopy, he is called a 'general of proper form.' Unless the general himself submits to these observances he will not have the means to know the cold and warmth of the officers and soldiers.

"If, when they advance into ravines and obstacles, or encounter muddy terrain, the general always takes the first steps, he is termed a 'general of strength.' If the general does not personally exert his strength, he has no means to know the labors and hardships of the officers and soldiers.

"If only after the men are settled in their encampment does the general retire; only after all the cooks have finished their cooking does he go in to eat; and if the army does not light fires to keep warm he also does not have one, he is termed a 'general who stifles desire.' Unless the general himself practices stifling his desires, he has no way to know the hunger and satiety of the officers and troops.

"The general shares heat and cold, labor and suffering, hunger and satiety with the officers and men. Therefore, when the masses of the Three Armies hear the sound of the drum they are happy, and when they hear the sound of the gong they are angry. When attacking a high wall or crossing a deep lake, under a hail of arrows and stones, the officers will compete to be first to scale the wall. When the naked blades clash, the officers will compete to be the first to go forward. It is not because they like death and take pleasure in being wounded, but because the general knows their feelings of heat and cold, hunger and satiety, and clearly displays his knowledge of their labor and suffering."

24 陰 符

Secret Tallies

KING WU ASKED THE T'AI KUNG: "If we lead the army deep into the territory of the feudal lords where the Three Armies suddenly suffer some delay or require urgent action, perhaps a situation to our advantage, or one to our disadvantage, and I want to communicate between those nearby and those more distant, to respond to the outside from the inside, in order to supply the use of the Three Armies—how should we do it?"

The T'ai Kung said: "The ruler and his generals have a system of secret tallies, altogether consisting of eight grades:

"There is a tally signifying a great victory over the enemy, one foot long.

"There is a tally for destroying the enemy's army and killing their general, nine inches long.

"There is a tally for forcing the surrender of the enemy's walls and capturing the town, eight inches long.

"There is a tally for driving the enemy back and reporting deep penetration, seven inches long.

"There is a tally to alert the masses to prepare for stalwart defensive measures, six inches long.

"There is a tally requesting supplies and additional soldiers, five inches long.

"There is a tally signifying the army's defeat and the general's death, four inches long.

"There is a tally signifying the loss of all advantages and the army's surrender, three inches long.

"Detain all those who bring in and present tallies, and if the information from the tally should leak out, execute all those who heard and told about it. These eight tallies, which only the ruler and general should secretly know, provide a technique for covert communication that will not allow outsiders to know the true situation. Accordingly, even though the enemy has the wisdom of a sage, no one will comprehend their significance."

King Wu said: "Excellent."

25 陰書

Secret Letters

KING WU ASKED THE T'AI KUNG: "The army has been led deep into the territory of the feudal lords, and the commanding general wants to bring the troops together, implement inexhaustible changes, and plan for unfathomable advantages. As these matters are quite numerous, a simple tally is inadequate to clearly express them. As they are separated by some distance, verbal communications cannot get through. What should we do?"

The T'ai Kung said: "Whenever you have secret affairs and major considerations, letters should be employed rather than tallies. The ruler sends a letter to the general; the general uses a letter to query the ruler. The letters are composed in one unit, then divided. They are sent out in three parts, with only one person knowing the contents.

'Divided' means it is separated into three parts. 'Sent out in three parts, with only one person knowing' means there are three messengers, each carrying one part, and when the three are compared together, only then does one know the contents. This is referred to as a 'secret letter.' Even if the enemy has the wisdom of a sage, they will not be able to recognize the contents."

"Excellent," said King Wu.

26 軍勢

The Army's Strategic Power

KING WU ASKED THE T'AI KUNG: "What is the Tao for aggressive warfare?"

The T'ai Kung replied: "Strategic power is exercised in accord with the enemy's movements. Changes stem from the confrontation between the two armies. Unorthodox and orthodox tactics are produced from the inexhaustible resources of the mind. Thus the greatest affairs are not discussed, and the employment of troops is not spoken about. Moreover, words that discuss ultimate affairs are not worth listening to. The employment of troops is not so definitive as to be visible. They go suddenly, they come suddenly. Only someone who can exercise sole control, without being governed by other men, is a military weapon.

"If your plans are heard about, the enemy will make counterplans. If you are perceived, they will plot against you. If you are known, they will put you in difficulty. If you are fathomed, they will endanger you.

"Thus one who excels in warfare does not await the deployment of forces. One who excels at eliminating the misfortunes of the people manages them before they appear. Conquering the enemy means being victorious over the formless. The superior fighter does not engage in battle. Thus one who fights and attains victory in front of naked blades is not a good general. One who makes preparations after the battle has been lost is not a Superior Sage! One whose wisdom is the same as the masses is not a general for the state. One whose skill is the same as the masses is not a State Artisan.

"In military affairs nothing is more important than certain victory. In employing the army nothing is more important than obscurity and silence. In movement nothing is more important than the unexpected. In planning nothing is more important than not being knowable.

"To be the first to gain victory, initially display some weakness to the enemy and only afterward do battle. Then your effort will be half, but the achievement will be doubled.

"The Sage takes his signs from the movements of Heaven and Earth; who knows his principles? He accords with the Tao of *yin* and *yang*, and follows their seasonal activity. He follows the cycles of fullness and emptiness of Heaven and Earth, taking them as his constant. All things have life and death in accord with the form of Heaven and Earth. Thus it is said that if one fights before seeing the situation, even if he is more numerous, he will certainly be defeated.

"One who excels at warfare will await events in the situation without making any movement. When he sees he can be victorious he will arise; if he sees he cannot be victorious he will desist. Thus it is said he does not have any fear, he does not vacillate. Of the many harms that can beset an army, vacillation is the greatest. Of disasters that can befall an army, none surpasses doubt.

"One who excels in warfare will not lose an advantage when he perceives it, nor be doubtful when he meets the moment. One who loses an advantage or lags behind the time for action will, on the contrary, suffer from disaster. Thus the wise follow the time and do not lose an advantage; the skillful are decisive and have no doubts. For this reason when there is a sudden clap of thunder there is not time to cover the ears; when there is a flash of lightning there is not time to close the eyes. Advance as if suddenly startled, employ your troops as if deranged. Those who oppose you will be destroyed, those who come near will perish. Who can defend against such an attack?

"Now when matters are not discussed and the general preserves their secrecy he is spiritlike. When things are not manifest but he discerns them he is enlightened. Thus if one knows the Tao of spirit and enlightenment, no enemies will act against him in the field, nor will any state stand against him."

"Excellent," said King Wu.

27 奇兵

The Unorthodox Army

KING WU ASKED THE T'AI KUNG: "In general, what are the great essentials in the art of employing the army?"

The T'ai Kung replied: "The ancients who excelled at warfare were not able to wage war above Heaven, nor could they wage war below Earth. Their success and defeat in all cases proceeded from the spiritual employment of strategic power. Those who attained it flourished; those who lost it perished.

"Now when our two armies, opposing each other, have deployed their armored soldiers and established their battle arrays, releasing some of your troops to create chaos in the ranks is the means by which to fabricate deceptive changes.

"Deep grass and dense growth are the means by which to effect a concealed escape.

"Valleys with streams and treacherous ravines are the means by which to stop chariots and defend against cavalry.

"Narrow passes and mountain forests are the means by which a few can attack a large force.

"Marshy depressions and secluded dark areas are the means by which to conceal your appearance.

"Deploying on clear, open ground without any concealment is the means by which to fight with strength and courage.

"Being as swift as a flying arrow, attacking as suddenly as the release of a crossbow, are the means by which to destroy brilliant plans.

"Setting up ingenious ambushes and preparing unorthodox troops, stretching out distant formations to deceive and entice the enemy, are the means by which to destroy the enemy's army and capture its general.

"Dividing your troops into four and splitting them into five are the means by which to attack their circular formations and destroy their square ones.

"Taking advantage of their fright and fear is the means by which one can attack ten.

"Taking advantage of their exhaustion and encamping at dusk is the means by which ten can attack a hundred.

"Unorthodox technical skills are the means by which to cross deep waters and ford rivers.

"Strong crossbows and long weapons are the means by which to fight across water.

"Distant observation posts and far-off scouts, explosive haste and feigned retreats are the means by which to force the surrender of walled fortifications and compel the submission of towns.

"Drumming an advance and setting up a great tumult are the means by which to implement unorthodox plans.

"High winds and heavy rain are the means by which to strike the front and seize the rear.

"Disguising some men as enemy emissaries is the means by which to sever their supply lines.

"Forging enemy commands and orders, and wearing the same clothes as the enemy, are the means by which to be prepared for their retreat.

"Warfare that is invariably in accord with righteousness is the means by which to incite the masses and be victorious over the enemy.

"Honored ranks and generous rewards are the means by which to encourage obeying orders.

"Severe punishments and heavy fines are the means by which to force the weary and indolent to advance.

"Happiness and anger, bestowing and taking away, civil and martial measures, at times slowly, at others rapidly—all these are the means by which to order and harmonize the Three Armies, to govern and unify subordinates.

"Occupying high ground is the means by which to be alert and assume a defensive posture.

"Holding defiles and narrows is the means by which to be solidly entrenched.

"Mountain forests and dense growth are the means by which to come and go silently.

"Deep moats, high ramparts, and large reserves of supplies are the means by which to sustain your position for a long time.

"Thus it is said: 'One who does not know how to plan for aggressive warfare cannot be spoken with about the enemy. One who cannot divide and move his troops about cannot be spoken with about unorthodox strategies. One who does not have a penetrating under-standing of both order and chaos cannot be spoken with about changes.'

"Accordingly it is said: 'If the general is not benevolent, then the Three Armies will not be close to him. If the general is not coura-geous, then the Three Armies will not be fierce. If the general is not wise, then the Three Armies will be greatly perplexed. If the general is not perspicacious, then the Three Armies will be confounded. If the general is not quick-witted and acute, then the Three Armies will lose the moment. If the general is not constantly alert, then the Three Armies will waste their preparations. If the general is not strong and forceful, then the Three Armies will fail in their duty.'

"Thus the general is their Master of Fate. The Three Armies are ordered with him, and they are disordered with him. If one obtains a Worthy to serve as general, the army will be strong and the state will prosper. If one does not obtain a Worthy as general, the army will be weak and the state will perish."

"Excellent," said King Wu.

28 五音

The Five Notes

KING WU ASKED THE T'AI KUNG: "From the sound of the pitch pipes can we know the fluctuations of the Three Armies, foretell victory and defeat?"

The T'ai Kung said: "Your question is profound indeed! Now there are twelve pipes, with five major notes: *kung, shang, chiao, cheng,* and *yü.* These are the true, orthodox sounds, unchanged for over ten thousand generations.

"The spirits of the five phases are constants of the Tao. Metal, wood, water, fire, and earth, each according to their conquest relationship, can be employed to attack the enemy. In antiquity, during the period of the Three Sage Emperors, they used the nature of vacuity and nonaction to govern the hard and strong. They did not have characters for writing; everything proceeded from the five phases. The Tao of the five phases is the naturalness of Heaven and Earth. The division into the six *chia* is a realization of marvelous and subtle spirit.

"Their method was, when the day had been clear and calm, without any clouds, wind, or rain, to send light cavalry out in the middle of the night to approach the enemy's fortifications. Stopping about nine hundred paces away they would all lift their pipes to their ears and then yell out to startle the enemy. There would be a very small, subtle sound that would respond in the pitch pipes.

"If the *chiao* note responded among the pipes, it indicated a White Tiger.

"If the *cheng* note responded in the pipes, it indicated the Mysterious Military.

"If the *shang* note responded in the pipes, it indicated the Vermilion Bird.

"If the *yü* note responded in the pipes, it indicated the Hooked Formation.

"If none of the five notes responded in the pipes, it was *kung*, signi-fying a Green Dragon.

"These signs of the five phases are evidence to assist in the con-quest, the subtle moments of success and defeat."

"Excellent," said King Wu.

The T'ai Kung continued: "These subtle, mysterious notes all have external indications."

"How can we know them?" King Wu asked.

The T'ai Kung replied: "When the enemy has been startled into movement listen for them. If you hear the sound of the *pao* drum, then it is *chiao*. If you see the flash of lights from a fire, then it is *cheng*. If you hear the sounds of bronze and iron, of spears and hal-berds, then it is *shang*. If you hear the sound of people sighing, it is *yü*. If all is silent, without any sound, then it is *kung*. These five are the signs of sound and appearance."

29 兵 徵

The Army's Indications

KING WU ASKED THE T'AI KUNG: "Before engaging in battle I want to first know the enemy's strengths and weaknesses, to foresee indications of victory or defeat. How can this be done?"

The T'ai Kung replied: "Indications of victory or defeat will be first manifest in their spirit. The enlightened general will investigate them, for they will be evidenced in the men.

"Clearly observe the enemy's coming and going, advancing and withdrawing. Investigate his movements and periods at rest, whether they speak about portents, what the officers and troops report. If the Three Armies are exhilarated; the officers and troops fear the laws; respect the general's commands; rejoice with each other in destroying the enemy; boast to each other about their courage and ferocity; and praise each other for their awesomeness and martial demeanor—these are indications of a strong enemy.

"If the Three Armies have been startled a number of times, the officers and troops no longer maintaining good order; they terrify each other with stories about the enemy's strength; they speak to each other about the disadvantages; they anxiously look about at each other, listening carefully; they talk incessantly of ill omens, a myriad of mouths confusing each other; they fear neither laws nor orders, and do not regard their general seriously—these are indications of weakness.

"When the Three Armies are well ordered; the deployment's strategic configuration of power solid, with deep moats and high ramparts; and moreover they enjoy the advantages of high winds and heavy rain, while the army is untroubled; the signal flags and pennants point to the front; the sound of the gongs and bells rises up and is clear; and the sound of the small and large drums clearly rises—these are indications of having obtained spiritual, enlightened assistance, foretelling a great victory.

"When their formations are not solid; their flags and pennants confused and entangled with each other; they go contrary to the advantages of high wind and heavy rain; their officers and troops are terrified; and their *ch'i* broken while they are not unified; their warhorses have been frightened and run off, their military chariots have broken axles; and the sound of their gongs and bells sinks down and is murky, while the sound of their drums is wet and damp—these are indications foretelling a great defeat.

"In general, when you attack city walls or surround towns, if the color of their *ch'i* is like dead ashes, the city can be slaughtered. If the city's *ch'i* drifts out to the north, the city can be conquered. If the city's *ch'i* goes out and drifts to the west, the city can be forced to surrender. If the city's *ch'i* goes out and drifts to the south, it cannot be taken. If the city's *ch'i* goes out and drifts to the east, the city cannot be attacked. If the city's *ch'i* goes out, but then drifts back in, the ruler has already fled. If the city's *ch'i* goes out and overspreads our army, the soldiers will surely fall ill. If the city's *ch'i* goes out and just rises up without any direction, the army will have to be employed for a long time. If you have attacked a walled city or surrounded a town for more than ten days without thunder or rain, you must hastily abandon it, for the city must have a source of great assistance. Those are the means by which to know that you can attack, and then go on to mount the attack, or that you should not attack, and therefore stop."

"Excellent," said King Wu.

30 農器

Agricultural Implements

KING WU ASKED THE T'AI KUNG, "If All under Heaven are at peace and settled, while the state is not engaged in any conflicts, can we dispense with maintaining the implements of war? Can we forego preparing equipment for defense?"

The T'ai Kung said: "The implements for offense and defense are fully found in ordinary human activity. Digging sticks serve as *chevaux-de-frise* and caltrops. Oxen and horse-pulled wagons can be used in the encampment and as covering shields. The different hoes can be used as spears and spear-tipped halberds. Raincoats of straw and large umbrellas serve as armor and protective shields. Large hoes, spades, axes, saws, mortars and pestles are tools for attacking walls. Oxen and horses are the means to transport provisions. Chickens and dogs serve as lookouts. The cloth that women weave serves as flags and pennants.

"The method that the men use for leveling the fields is the same for attacking walls. The skill needed in spring to cut down grass and thickets is the same as needed for fighting against chariots and cavalry. The weeding methods used in summer are the same as used in battle against foot soldiers. The grain harvested and the firewood cut in the fall will be provisions for the military. In the winter well-filled granaries and storehouses will ensure a solid defense.

"The units of five found in the fields and villages will provide the tallies and good faith that bind the men together. The villages have officials and the offices have chiefs who can lead the army. The villages have walls surrounding them that are not crossed; they provide the basis for the division into platoons. The transportation of grain and the cutting of hay provides for the state storehouses and armories. The skills used in repairing the inner and outer walls in the spring and fall, in maintaining the moats and channels, are used to build ramparts and fortifications.

"Thus the tools for employing the military are completely found in ordinary human activity. One who is good at governing a state will take them from ordinary human affairs. Then they must be made to accord with the good management of the six animals; the opening up of wild lands; and the settling of the people where they dwell. The husband has a number of acres that he farms, the wife a measured amount of material to weave—this is the Way to enrich the state and strengthen the army."

"Excellent," said King Wu.

IV 虎韜

TIGER
SECRET
TEACHING

31 軍用

The Army's Equipment

KING WU ASKED THE T'AI KUNG: "When the king mobilizes the Three Armies, are there any rules for determining the army's equipment, such as the implements for attack and defense, including type and quantity?"

The T'ai Kung said: "A great question, my King! The implements for attack and defense each have their own categories. This results in the great awesomeness of the army."

King Wu said: "I would like to hear about them."

The T'ai Kung replied: "As for the basic numbers when employing the army, if commanding ten thousand armed soldiers, the rules for the various types of equipment and their employment are as follows.

"Thirty-six Martial-protective Large Chariots. Skilled officers, strong crossbowmen, spear-bearers, and halberdiers, a total of twenty-four, for each flank and the rear. The chariots have eight-foot wheels. On them are set up pennants and drums that, according to *The Art of War*, are referred to as 'Shaking Fear.' They are used to penetrate solid formations, to defeat strong enemies.

"Seventy-two Martial-flanking Large Covered Spear-and-Halberd Chariots. Skilled officers, strong crossbowmen, spearbearers, and halberdiers comprise the flanks. They have five-foot wheels and winch-powered, linked crossbows that fire multiple arrows for self-protection. They are used to penetrate solid formations and defeat strong enemies.

"One hundred and forty Flank-supporting Small Covered Chariots equipped with winch-powered, linked crossbows to fire multiple arrows for self-protection. They have deer wheels, and are used to penetrate solid formations and defeat strong enemies.

"Thirty-six Great Yellow Triple-linked Crossbow Large Chariots. Skilled officers, strong crossbowmen, spearbearers, and halberdiers comprise the flanks, with 'flying duck' and 'lightning's shadow' arrows for self-protection. Flying duck arrows have red shafts and white feathers, with bronze arrowheads. Lightning's shadow arrows have green shafts and red feathers, with iron heads. In the daytime they display pennants of red silk, six feet long by six inches wide, that shimmer in the light. At night they hang pennants of white silk, also six feet long by six inches wide, that appear like meteors. They are used to penetrate solid formations, to defeat infantry and cavalry.

"Thirty-six Great Attack Chariots. Carrying Praying Mantis Martial Knights, they can attack both horizontal and vertical formations, and can defeat the enemy.

"Baggage Chariots for repelling mounted invaders, also called 'Lightning Chariots.' *The Art of War* refers to their use in 'lightning attacks.' They are used to penetrate solid formations, to defeat both infantry and cavalry.

"One hundred and sixty Spear-and-Halberd Light Chariots for repelling night invaders from the fore. Each carries three Praying Mantis Martial Knights. *The Art of War* refers to them as mounting 'thunder attacks.' They are used to penetrate solid formations, to defeat both infantry and cavalry.

"Iron truncheons with large, square heads weighing twelve catties, and shafts more than five feet long, twelve hundred of them. Also termed 'Heaven's Truncheon.'

"The Great Handle *Fu* Ax, with an eight-inch blade, weighing eight catties, and a shaft more than five feet long, twelve hundred of them. Also termed 'Heaven's *Yüeh* Ax.'"

"Also the Iron Square-Headed Pounder, weighing eight catties, with a shaft of more than five feet, twelve hundred. Also termed 'Heaven's Pounder.' They are used to defeat infantry and hordes of mounted invaders.

"The Flying Hook, eight inches long. The curve of the hook is five inches long, the shaft is more than six feet long. Twelve hundred of them. They are thrown into masses of soldiers.

"To defend the Three Armies deploy chariots equipped with

wooden Praying Mantises and sword blades, each twenty feet across, altogether one hundred and twenty of them. They are also termed *chevaux-de-frise*. On open, level ground the infantry can use them to defeat chariots and cavalry.

"Wooden caltrops that stick out of the ground about two feet five inches, one hundred twenty. They are employed to defeat infantry and cavalry, to urgently press the attack against invaders, and intercept their flight.

"Short-axle Quick-turning Spear-and-Halberd Chariots, one hundred twenty. They were employed by the Yellow Emperor to vanquish Ch'ih Yu. They are used to defeat both infantry and cavalry, to urgently press the attack against the invaders, and intercept their flight.

"For narrow roads and small bypaths, set out iron caltrops eight inches wide, having hooks four inches high and shafts of more than six feet, twelve hundred. They are for defeating retreating cavalry.

"If, in the darkness of night the enemy should suddenly press an attack and the naked blades clash, stretch out a ground net, and spread out two arrowheaded caltrops connected together with 'weaving women' type caltrops on both sides. The points of the blades should be about two feet apart. Twelve thousand sets.

"For fighting in wild expanses and in the middle of tall grass there are the square-shank, arrow-shaped spears, twelve hundred of them. The method for deploying these spears is to have them stick out of the ground one foot five inches. They are used to defeat infantry and cavalry, to urgently press the attack against invaders, and intercept their flight.

"On narrow roads, small bypaths, and constricted terrain set out iron chains, one hundred twenty of them, to defeat infantry and cavalry, urgently press the attack against the invaders, and intercept their flight.

"For the protection and defense of the gates to fortifications there are the small mobile shields with spear and halberd tips affixed, twelve of them, and winch-driven, multiple-arrow crossbows for self-protection.

"For the protection of the Three Armies there are Heaven's Net and Tiger's Drop, linked together with chains, one hundred twenty of them. One array is fifteen feet wide and eight feet tall. For the chariots with Tiger's Drop and sword blades affixed, the array is fifteen feet wide, and eight feet tall. Five hundred ten of them.

"For crossing over moats and ditches there is the Flying Bridge. One section is fifteen feet wide and more than twenty feet long. Eight of them. On top there are swivel winches to extend them by linked chains.

"For crossing over large bodies of water there is the Flying River, eight of them. They are fifteen feet wide and more than twenty feet long, and are extended by linked chains.

"There is also the Heavenly Float with Iron Praying Mantis, rectangular inside, circular outside, four feet or more in diameter, equipped with plantern winches. Thirty-two of them. When the Heavenly Floats are used to deploy the Flying River to cross a large lake they are referred to as 'Heaven's Huang,' and also termed 'Heaven's Boat.'

When in mountain forests or occupying the wilds, connect the Tiger's Drops to make a fenced encampment. Employ iron chains, length of more than twenty feet, twelve hundred sets. Also employ large ropes with rings, girth of four inches, length of more than forty feet, six hundred; mid-sized ropes with rings, girth of two inches, length of forty feet or more, two hundred sets; and small, braided cords with rings, length of twenty feet or more, twelve thousand.

"Wooden canopies for covering the heavy chariots called 'Heaven's Rain,' which fit together along serrated seams, each four feet wide and more than four feet long, one for each chariot. They are erected by using small iron posts.

"For cutting trees there is the Heavenly Ax, which weighs eight catties. Its handle is more than three feet long. Three hundred of them. Also the mattock with a blade six inches wide and a shaft more than five feet long, three hundred.

"Copper rams for pounding, more than five feet long, three hundred.

"Eagle claws, with square hafts, iron handles, and shafts more than seven feet long, three hundred.

"Square-shafted iron pitchforks with handles more than seven feet long, three hundred.

"Square-shafted double pronged iron pitchforks with shafts more than seven feet long, three hundred.

"Large sickles for cutting grass and light trees, with shafts more than seven feet long, three hundred.

"Great oar-shaped blades, weight of eight catties, with shafts more than six feet long, three hundred.

"Iron stakes with rings affixed at top, more than three feet long, three hundred.

"Large hammers for pounding posts, weight of five catties, handles more than two feet long, one hundred twenty.

"Armored soldiers, ten thousand. Strong crossbowmen, six thousand. Halberdiers with shields, two thousand. Spearmen with shields, two thousand. Skilled men to repair offensive weapons and sharpen them, three hundred.

"These then are the general numbers required for each category when raising an army."

King Wu said: "I accept your instructions."

32 三陣

Three Deployments

KING WU ASKED THE T'AI KUNG: "In employing the army there are the Heavenly Deployment, the Earthly Deployment, and the Human Deployment. What are these?"

The T'ai Kung replied: "When you accord with the sun and moon, the stars, planets, and the handle of the Big Dipper, one on the left, one on the right, one in front, and one to the rear, this is referred to as the Heavenly Deployment.

"When the hills and mounds, rivers and streams are similarly to your advantage to the front, rear, left, and right, this is referred to as the Earthly Deployment.

"When you employ chariots and horses, when you use both the civil and martial, this is referred to as the Human Deployment."

"Excellent," said King Wu.

33 疾 戰

Urgent Battles

KING WU ASKED THE T'AI KUNG: "If the enemy surrounds us, severing both our advance and retreat, breaking off our supply lines, what should we do?"

The T'ai Kung said: "These are the most distressed troops in the world! If you employ them explosively you will be victorious; if you are slow to employ them, you will be defeated. In this situation, if you deploy your troops into martial assault formations on the four sides; use your military chariots and valiant cavalry to startle and confuse their army; and urgently attack them, you can thrust across them."

King Wu asked: "After we have broken out of the encirclement, if we want to take advantage of it to gain victory, what should we do?"

The T'ai Kung said: "The Army of the Left should urgently strike out to the left, and the Army of the Right should urgently strike out to the right. But do not get entangled in protracted fighting with the enemy over any one road. The Central Army should alternately move to the front and then the rear. Even though the enemy is more numerous, their general can be driven off."

34 必 出

Certain Escape

KING WU ASKED THE T'AI KUNG: "Suppose we have led our troops deep into the territory of the feudal lords where the enemy unites from all quarters and surrounds us, cutting off our road back home, and severing our supply lines. The enemy is numerous and extremely well provisioned, while the ravines and gorges are also solidly held. We must get out—how can we?"

The T'ai Kung said: "In the matter of effecting a certain escape your equipment is your treasure, while courageous fighting is foremost. If you investigate and learn where the enemy's terrain is empty and vacuous, the places where there are no men, you can effect a certain escape.

"Order your generals and officers to carry the Mysterious Dark Pennants and take up the implements of war. Require the soldiers to put wooden gags into their mouths. Then move out at night. Men of courage, strength, and swiftness, who will risk danger, should occupy the front, to level fortifications and open a passage for the army. Skilled soldiers and strong crossbowmen should compose an ambushing force, which will remain in the rear. Your weak soldiers, chariots, and cavalry should occupy the middle. When the deployment is complete slowly advance, being very cautious not to startle or frighten the enemy. Have the Martial Attack Chariots defend the front and rear, and the Martial Flanking Great Covered Chariots protect the left and right flanks.

"If the enemy should be startled, have your courageous, strong risk-takers fervently attack and advance. The weaker troops, chariots, and cavalry should bring up the rear. Your skilled soldiers and strong crossbowmen should conceal themselves in ambush. If you determine that the enemy is in pursuit, the men lying in ambush should swiftly attack their rear. Make your fires and drums numerous, and attack as if coming out of the very ground, or dropping from Heaven above. If the Three Armies fight courageously no one will be able to withstand us!"

King Wu said: "In front of us lies a large body of water, or broad moat, or deep water hole that we want to cross. However, we do not have equipment such as boats and oars. The enemy has fortifications and ramparts that limit our army's advance and block off our retreat. Patrols are constantly watchful, passes fully defended. Their chariots and cavalry press us in front, their courageous fighters attack us to the rear. What should we do?"

The T'ai Kung said: "Large bodies of water, broad moats, and deep water holes are usually not defended by the enemy. If they are able to defend them, their troops will certainly be few. In such situations you should use the Flying River with winches, and also the Heavenly Huang, to cross the army over. Our courageous, strong, skilled soldiers should move where we indicate, rushing into the enemy, breaking up their formations, all fighting to the death.

"First of all, burn the supply wagons and provisions, and clearly inform the men that those who fight courageously will live, while cowards will die. After they have broken out and crossed the bridges, order the rear elements to set a great conflagration visible from far off. The troops sallying forth must take advantage of the cover afforded by grass, trees, hillocks, and ravines. The enemy's chariots and cavalry will certainly not dare pursue them too far. Using the flames as a marker, the first to go out should be ordered to proceed as far as the flames and then stop, reforming a four-sided attack formation. In this fashion the Three Armies will be fervent and sharp, and fight courageously, and no one will be able to withstand us."

King Wu said: "Excellent!"

35 軍畧

Planning for the Army

KING WU ASKED THE T'AI KUNG: "Suppose we have led the army deep into the territory of the feudal lords where we encounter deep streams, or water in large valleys, ravines, and defiles. Our Three Armies have not yet fully forded them when Heaven lets loose a torrent, resulting in a sudden flood surge. The rear cannot maintain contact with the advance portion. We do not have equipment such as pontoon bridges, nor materials such as heavy grass to stem the waters. I want to finish crossing, to keep the Three Armies from becoming bogged down. What should I do?"

The T'ai Kung said: "If the leader of the army and commander of the masses does not first establish his plans, the proper equipment will not be prepared. If his instructions are not precise and trusted, the officers and men will not be trained. Under such conditions they cannot comprise a king's army.

"In general, when the army is involved in a major campaign everyone should be trained to use the equipment. For attacking a city wall or surrounding a town there are armored assault chariots, overlook carts, and battering rams, while for seeing inside the walls there are 'cloud ladders' and 'flying towers.' If the advance of the Three Armies is stopped, then there are the Martial Assault Great Chariots. For defending both front and rear, for severing roads and blocking streets, there are the skilled soldiers and strong crossbowmen who protect the two flanks. If you are encamping or building fortifications, there are the Heaven's Net, the Martial Drop, the *chevaux-de-frise*, and the caltrops.

"In the daytime climb the cloud ladder and look off into the distance. Set up five-colored pennants and flags. At night set out ten thousand fire-cloud torches, beat the thunder drums, strike the war drums and bells, and blow the sharp sounding whistles.

"For crossing over moats and ditches there are Flying Bridges with plantern-mounted winches and cogs. For crossing large bodies of water there are boats called the Heavenly Huang and Flying River. For going against the waves and up current there are the Floating Ocean rafts and the rope-pulled River Severance. When the equipment to be used by the Three Armies is fully prepared, what worries will the commander-in-chief have?"

36 臨 境

Approaching the Border

KING WU ASKED THE T'AI KUNG: "Both the enemy and our army have reached the border where we are in a standoff. They can approach, and we can also advance. Both deployments are solid and stable; neither side dares to move first. We want to go forth and attack them, but they can also come forward. What should we do?"

The T'ai Kung said: "Divide the army into three sections. Have our advance troops deepen the moats and increase the height of the ramparts, but none of the soldiers should go forth. Array the flags and pennants, beat the leather war drums, and complete all the defensive measures. Order our rear army to stockpile supplies and foodstuffs, without causing the enemy to know our intentions. Then send forth our elite troops to secretly launch a sudden attack against their center, striking where they do not expect it, attacking where they are not prepared. Since the enemy does not know our real situation, they will stop and not advance."

King Wu asked: "Suppose the enemy knows our real situation and has fathomed our plans. If we move, they will be able to learn everything about us. Their elite troops are concealed in the deep grass. They press us on the narrow roads, and are attacking where convenient for them. What should we do?"

The T'ai Kung said: "Every day have the vanguard go forth and instigate skirmishes with them in order to psychologically wear them out. Have our older and weaker soldiers drag brushwood to stir up the dust, beat the drums and shout, and move back and forth—some going to the left, some to the right, never getting closer than a hundred paces from the enemy. Their general will certainly become fatigued, and their troops will become fearful. In this situation the enemy will not dare come forward. Then our advancing troops should unexpectedly not stop, some continuing forward to attack their interior, others the exterior. With our Three Armies all fervently engaging in the battle, the enemy will certainly be defeated."

37 動 靜

Movement and Rest

KING WU ASKED THE T'AI KUNG: "Suppose we have led our troops deep into the territory of the feudal lords and are confronting the enemy. The two deployments looking across at each other are equal in numbers and strength, and neither dares to move first. I want to cause the enemy's general to become terrified; their officers and men to become dispirited; their battle array to become unstable; their reserve army to want to run off; and those deployed forward to constantly look about at each other. I want to beat the drums, set up a clamor, and take advantage of it so that the enemy will then run off. How can we do it?"

The T'ai Kung said: "In this case send our troops out about ten kilometers from the enemy and have them conceal themselves on both flanks. Send your chariots and cavalry out about one hundred kilometers and have them return unobserved to assume positions cutting across both their front and rear. Multiply the number of flags and pennants, and increase the number of gongs and drums. When the battle is joined, beat the drums, set up a clamor, and have your men all rise up together. The enemy's general will surely be afraid, and their army will be terrified. Large and small numbers will not come to each other's rescue; upper and lower ranks will not wait for each other; and the enemy will definitely be defeated."

King Wu asked: "Suppose because of the enemy's strategic configuration of power we cannot conceal troops on the flanks, and moreover our chariots and cavalry have no way to cross behind them and assume positions both to the front and rear. The enemy anticipates my thoughts and makes preemptive preparations. Our officers and soldiers are dejected, our generals are afraid. If we engage in battle we will not be victorious. What then?"

The T'ai Kung said: "Truly a serious question. In this case five days before engaging in battle dispatch distant patrols to observe their activities and analyze their forward movement in order to prepare an ambush and await them. We must meet the enemy on deadly ground. Spread our flags and pennants out over a great distance, disperse our arrays and formations. We must race forward to meet the enemy. After the battle has been joined, suddenly retreat, beating the gongs incessantly. Withdraw about three kilometers beyond the ambush, then turn about and attack. Your concealed troops should simultaneously arise. Some should penetrate the flanks, others attack their vanguard and rear-guard positions. If the Three Armies fervently engage in battle, the enemy will certainly run off."

King Wu said: "Excellent."

38 金鼓

Gongs and Drums

KING WU ASKED THE T'AI KUNG: "Suppose we have led the army deep into the territory of the feudal lords where we are confronting the enemy. The weather has been either extremely hot or very cold, and day and night it has been raining incessantly for ten days. The ditches and ramparts are all collapsing; defiles and barricades are unguarded; our patrols have become negligent; and the officers and men are not alert. Suppose the enemy comes at night. Our Three Armies are unprepared, while the upper and lower ranks are confused and disordered. What should we do?"

The T'ai Kung said: "In general, for the Three Armies alertness makes for solidity, while laziness results in defeat. Order our guards on the ramparts to unceasingly challenge everyone. Have all those bearing the signal flags, both inside and outside the encampment, watch each other, responding to each other's orders with countersigns, but do not allow them to make any noise. All efforts should be externally oriented.

"Three thousand men should comprise a detachment. Instruct and constrain them with an oath, requiring each of them to exercise vigilance at his post. If the enemy approaches, when they see our state of readiness and alertness they will certainly turn around. As a result their strength will become exhausted and their spirits dejected. At that moment send forth our elite troops to follow and attack them."

King Wu asked: "The enemy, knowing we are following them, conceals elite troops in ambush while pretending to continue to retreat. When we reach the ambush their troops turn back, some attacking our front, others our rear, while some press our fortifications. Our Three Armies are terrified, and in confusion fall out of formation and leave their assigned positions. What should we do?"

The T'ai Kung said: "Divide into three forces, then follow and pursue them, but do not cross beyond their ambush. When all three forces have arrived, some should attack the front and rear, others should penetrate the two flanks. Make your commands clear, choose your orders carefully. Fervently attack, advancing forward, and the enemy will certainly be defeated."

39 絕道

Severed Routes

KING WU ASKED THE T'AI KUNG: "Suppose we have led the army deep into the territory of the feudal lords where, confronting them, we have each assumed defensive positions. The enemy has severed our supply routes, and occupied positions cutting across both our front and rear. If I want to engage them in battle we cannot win; but if I want to maintain our position we cannot hold out for long. What should we do?"

The T'ai Kung said: "In general, when you venture deep beyond the enemy's borders you must investigate the configuration and strategic advantages of the terrain, and concentrate upon seeking out and improving the advantages. Rely upon mountains, forests, ravines, rivers, streams, woods, and trees to secure defense. Carefully guard passes and bridges, and moreover be certain you know the advantages of terrain conveyed by the various cities, towns, hills, and funeral mounds. In this way the army will be solidly entrenched. The enemy will not be able to sever our supply routes, nor be able to occupy positions cutting across our front and rear."

King Wu asked: "Suppose, after our Three Armies have passed through a large forest or across a broad marsh and are on flat, accessible terrain, due to some erroneous or lost signal from our scouts, the enemy suddenly falls upon us. If we engage them in battle we cannot win; if we assume a defensive position it will not be secure. The enemy has outflanked us on both sides, and occupied positions cutting across our front and rear. The Three Armies are terrified. What should be done?"

The T'ai Kung said: "Now the rule for commanding an army is always to first dispatch scouts far forward so that when you are two hundred kilometers from the enemy you will already know their location. If the strategic configuration of the terrain is not advantageous, then use the Martial Attack chariots to form a mobile rampart and advance. Also establish two rear-guard armies to the rear, the further a one hundred kilometers away, the nearer fifty kilometers away. Thus, when there is a sudden alarm or urgent situation, both front and rear will know about it, and the Three Armies will always be able to complete their deployment into a solid formation, never suffering any destruction or harm."

"Excellent," said King Wu.

40 暑地

Occupying Enemy Territory

KING WU ASKED THE T'AI KUNG: "Suppose, being victorious in battle, we have deeply penetrated the enemy's territory and occupy his land. However, large walled cities remain that cannot be subjugated, while their second army holds the defiles and ravines, standing off against us. We want to attack the cities and besiege the towns, but I am afraid that their second army will suddenly appear and strike us. If their forces inside and outside unite in this fashion, they will oppose us from both within and without. Our Three Armies will be in chaos, the upper and lower ranks will be terrified. What should be done?"

The T'ai Kung said: "In general, when attacking cities and besieging towns, the chariots and cavalry must be kept at a distance. The encamped and defensive units must be on constant alert in order to obstruct the enemy both within and without. When the inhabitants have their food cut off, those outside being unable to transport anything in to them, those within the city walls will be afraid, and their general will certainly surrender."

King Wu said: "Suppose that when the supplies inside the city are cut off, external forces being unable to transport anything in, they clandestinely make a covenant and take an oath, concoct secret plans, and then sally forth at night, throwing all their forces into a death struggle. Some of their chariots, cavalry, and elite troops assault us from within, others attack from without. The officers and troops are confused, the Three Armies defeated and in chaos. What should be done?"

The T'ai Kung said: "In this case you should divide your forces into three armies. Be careful to evaluate the terrain's configuration and then strategically emplace them. You must know in detail the location of the enemy's second army, as well as his large cities and secondary fortifications. Leave them a passage in order to entice them to flee. Pay attention to all the preparations, not neglecting anything. The enemy will be afraid, and if they do not enter the mountains or the forests, they will return to the large towns, or run off to join the second army. When their chariots and cavalry are far off, attack the front, do not allow them to escape. Since those remaining in the city will think that the first to go out have a direct escape route, their well-trained troops and skilled officers will certainly issue forth, with the old and weak alone remaining. When our chariots and cavalry have deeply penetrated their territory, racing far off, none of the enemy's army will dare approach. Be careful not to engage them in battle; just sever their supply routes, surround and guard them, and you will certainly outlast them.

"Do not set fire to what the people have accumulated, do not destroy their palaces or houses, nor cut down the trees at grave sites or altars. Do not kill those who surrender, nor slay your captives. Instead show them benevolence and righteousness, extend your generous Virtue to them. Cause their people to say 'the guilt lies with one man.' In this way the entire realm will then submit."

"Excellent," said King Wu.

41 火戰

Incendiary Warfare

KING WU ASKED THE T'AI KUNG: "Suppose we have led our troops deep into the territory of the feudal lords where we encounter deep grass and heavy growth that surround our army on all sides. The Three Armies have traveled several hundred kilometers, the men and horses are exhausted and have halted to rest. Taking advantage of the extremely dry weather and a strong wind, the enemy ignites fires upwind from us. Their chariots, cavalry, and elite forces are firmly concealed in ambush to our rear. The Three Armies become terrified, scatter in confusion, and run off. What can be done?"

The T'ai Kung said: "Under such circumstances use the cloud ladders and flying towers to look far out to the left and right, to carefully investigate front and rear. When you see the fires arise, then set fires in front of our own forces, spreading them out over the area. Also set fires to the rear. If the enemy comes, withdraw the army and take up entrenched positions on the blackened earth to await their assault. In the same way, if you see flames arise to the rear, you must move far away. If we occupy the blackened ground with our strong crossbowmen and skilled soldiers protecting the left and right flanks, we can also set fires to the front and rear. In this way the enemy will not be able to harm us."

King Wu asked: "Suppose the enemy has set fires to the left and right, and also to the front and rear. Smoke covers our army, while his main force appears from over the blackened ground. What should we do?"

The T'ai Kung said: "In this case, assuming you have prepared a burnt section of ground, disperse the Martial Attack chariots to form a fighting barrier on all four sides, and have strong crossbowmen cover the flanks. This method will not bring victory, but will also not end in defeat."

42

Empty Fortifications

KING WU ASKED THE T'AI KUNG: "How can I know whether the enemy's fortifications are empty or full, whether their army is coming or going?"

The T'ai Kung said: "A general must know the Tao of Heaven above, the advantages of Earth below, and human affairs in the middle. You should mount high and look out far in order to see the enemy's changes and movements. Observe his fortifications, and then you will know whether they are empty or full. Observe his officers and troops, and then you will know whether they are coming or going."

King Wu asked: "How will I know it?"

The T'ai Kung said: "Listen to see if his drums are silent, if his bells make no sound. Look to see whether there are many birds flying above the fortifications, if they were not startled into flight. If there are not any vapors overhead you will certainly know the enemy has tricked you with dummies.

"If enemy forces precipitously go off, but not very far, and then return before assuming proper formation, they are using their officers and men too quickly. When they act too quickly, the forward and rear are unable to maintain good order. When they cannot maintain good order, the entire battle disposition will be in chaos. In such circumstances quickly dispatch troops to attack them. If you use a small number to strike a large force, they will certainly be defeated."

V 豹韜

LEOPARD
SECRET
TEACHING

43 林戰

Forest Warfare

KING WU ASKED THE T'AI KUNG: "Suppose we have led our troops deep into the territory of the feudal lords where we encounter a large forest that we share with the enemy in a standoff. If we assume a defensive posture I want it to be solid, or if we fight, to be victorious. How should we proceed?"

The T'ai Kung said: "Have our Three Armies divide into the assault formation. Improve the positions the troops will occupy, and station the archers and crossbowmen outside, with those carrying spear-tipped halberds and shields inside. Cut down and clear away the grass and trees, and extensively broaden the passages in order to facilitate our deployment onto the battle site. Set our pennants and flags out on high, and carefully encourage the Three Armies without letting the enemy know our true situation. This is referred to as 'Forest Warfare.'

"The method of Forest Warfare is to form the spearbearers and halberdiers into squads of five. If the woods are not dense, cavalry can be used in support. Battle chariots will occupy the front. When opportune they will fight; when not opportune they will desist. Where there are numerous ravines and defiles in the forest, you must deploy into the Assault Formation in order to be prepared both front and rear. If the Three Armies urgently attack, even though the enemy is numerous, they can be driven off. The men should fight and rest in turn, each with their sections. This is the main outline of Forest Warfare."

44 突戰

Explosive Warfare

KING WU ASKED THE T'AI KUNG: "Suppose the enemy's advance forces have penetrated deep into our territory and are ranging widely, occupying our land and driving off our cattle and horses. Then their Three Armies arrive en masse and press us outside our city walls. Our officers and troops are sorely afraid, our people are in bonds, having been captured by the enemy. If we assume a defensive posture, I want it to be solid; or if we fight, to be victorious. What should we do?"

The T'ai Kung said: "An enemy in situations such as this is referred to as an 'Explosive Force.' Their oxen and horses will certainly not have been fed, their officers and troops will have broken their supply routes, having explosively attacked and advanced. Order our distant towns and other armies to select their elite soldiers and urgently strike their rear. Carefully consult the calendar, for we must unite on a moonless night. The Three Armies should fight intensely, for then even though the enemy is numerous, their general can be captured."

King Wu said: "Suppose the enemy divides their forces into three or four detachments, some fighting with us and occupying our territory, others stopping to round up our oxen and horses. Their main army has not yet completely arrived, but they have had their swift invaders press us below the city walls. Therefore our Three Armies are sorely afraid. What should we do?"

The T'ai Kung said: "Carefully observe the enemy. Before they have all arrived, make preparations and await them. Go out about four kilometers from the walls and establish fortifications, setting out in good order our gongs and drums, flags and pennants. Our other troops will comprise an ambushing force. Order large numbers of strong crossbowmen to the top of the fortifications. Every hundred paces set up an 'explosive gate,' outside of which we should place the *chevaux-de-frise*. Our chariots and cavalry should be held outside, while our courageous, strong, fierce fighters should be secreted in this outer area. If the enemy should reach us, have our light-armored foot soldiers engage them in battle, then feign a retreat. Have the forces on top of the city wall set out the flags and pennants and strike the war drums, completing all preparations to defend the city. The enemy will assume we are going to defend the wall, and will certainly press an attack below it. Then release the forces lying in ambush, some to assault their interior, others to strike the exterior. Then the Three Armies should urgently press the attack, some striking the front lines, others the rear. Even their courageous soldiers will not be able to fight, while the swiftest will not have time to flee. This is termed 'Explosive Warfare.' Although the enemy is numerically superior, they will certainly run off."

"Excellent," said King Wu.

45 敵彊

Strong Enemy

KING WU ASKED THE T'AI KUNG: "Suppose we have led the army deep into the territory of the feudal lords until we are opposed by the enemy's assault forces. The enemy is numerous while we are few. The enemy is strong while we are weak. The enemy approaches at night, some attacking the left, others the right. The Three Armies are quaking. We want to be victorious if we choose to fight, and solid if we choose to maintain a defensive posture. How should we act?"

The T'ai Kung said: "In this case we refer to them as 'Shaking Invaders.' It is more advantageous to go out and fight; you cannot be defensive. Select skilled soldiers and strong crossbowmen, together with chariots and cavalry, to comprise the right and left flanks. Then urgently strike their forward forces, quickly attacking the rear as well.

Some should strike the exterior, others the interior. Their troops will certainly be confused, their generals afraid."

King Wu asked: "Suppose the enemy has blocked off our forward units some distance away, and is pressing a fervent attack on our rear. He has broken up our elite troops, and cut off our skilled soldiers. Our interior and exterior forces cannot communicate with each other. The Three Armies are in chaos, all running off in defeat. The officers and troops have no will to fight, the generals and commanders no desire to defend themselves. What should we do?"

The T'ai Kung said: "Illustrious is your question, my King! You should make your commands clear and be careful about your orders. You should have our courageous, crack troops who are willing to confront danger sally forth, each man carrying a torch, two men to a drum. You must know the enemy's location, then strike both the interior and exterior. When our secret signals have all been communicated, order them to extinguish the torches, and stop beating all the drums. Then the interior and exterior should respond to each other, each according to the appropriate time. When our Three Armies urgently attack, the enemy will certainly be defeated and vanquished."

"Excellent," said King Wu.

46 敵 武

Martial Enemy

KING WU ASKED THE T'AI KUNG: "Suppose we have led the army deep into the territory of the feudal lords where we suddenly encounter a martial, numerically superior enemy. If his martial chariots and valiant cavalry attack our left and right flanks, and our Three Armies become so shaken that their flight is unstoppable, what should I do?"

The T'ai Kung said: "In this situation you have what is termed a 'defeated army.' Those who are skillful in employing their forces will manage a victory. Those who are not will perish."

King Wu asked: "What does one do?"

The T'ai Kung replied: "Have our most skilled soldiers and strong crossbowmen, together with our martial chariots and valiant cavalry, conceal themselves on both sides of the retreat route, about three kilometers ahead and behind the main force. When the enemy pursues us, launch a simultaneous chariot and cavalry assault from both sides. In such circumstances the enemy will be thrown into confusion, and our fleeing soldiers will stop by themselves."

King Wu continued: "Suppose the enemy's chariots and cavalry are squarely opposite ours, but the enemy is numerous while we are few, the enemy strong while we are weak. Their approach is disciplined and spirited, and our formations are unable to withstand them. What should we do?"

The T'ai Kung replied: "Select our skilled soldiers and strong crossbowmen, and have them lie in ambush on both sides, while the chariots and cavalry deploy into a solid formation and assume position. When the enemy passes our concealed forces, the crossbowmen should fire en masse into their flanks. The chariots, cavalry, and skilled soldiers should then urgently attack their army, some striking the front, others striking the rear. Even if the enemy is numerous they will certainly flee."

"Excellent," said King Wu.

47

Crow and Cloud Formation in the Mountains

KING WU ASKED THE T'AI KUNG: "Suppose we have led the army deep into the territory of the feudal lords where we encounter high mountains with large flat rock outcroppings, on top of which are numerous peaks, all devoid of grass and trees. We are surrounded on all four sides by the enemy. Our Three Armies are afraid, the officers and troops confused. I want to be solid if we choose to defend our position, and victorious if we fight. What should we do?"

The T'ai Kung said: "Whenever the Three Armies occupy the heights of a mountain they are trapped on high by the enemy. When they hold the land below the mountain they are imprisoned by the forces above them. If you have already occupied the top of the mountain, you must prepare the Crow and Cloud Formation. The Crow and Cloud Formation should be prepared on both the *yin* and *yang* sides of the mountain. Some will encamp on the *yin* side, others will encamp on the *yang* side. Those that occupy the *yang* side must prepare against attacks from the *yin* side. Those occupying the *yin* side must prepare against attacks from the *yang* side. Those occupying the left side of the mountain must prepare against the right side. Those on the right, against the left. Wherever the enemy can ascend the mountain, your troops should establish external lines. If there are roads passing through the valley, sever them with your war chariots. Set your flags and pennants up high. Be cautious in commanding the Three Armies, do not allow the enemy to know your true situation. This is referred to as a 'mountain wall.'

"After your lines have been set, your officers and troops deployed, rules and orders already issued, and tactics, both orthodox and unorthodox, already planned, deploy your assault formation at the outer perimeter of the mountain, and have them improve the positions they occupy. Thereafter divide your chariots and cavalry into the Crow and Cloud Formation. When your Three Armies urgently attack the enemy, even though the latter are numerous, their general can be captured."

48 烏雲澤兵

Crow and Cloud Formation in the Marshes

KING WU ASKED THE T'AI KUNG: "Suppose we have led the army deep into the territory of the feudal lords where we are confronting the enemy across a river. The enemy is well equipped and numerous; we are impoverished and few. If we cross the water to attack we will not be able to advance, while if we want to outlast them our supplies are too few. We are encamped on salty ground. There are no towns in any direction, and moreover no grass or trees. There is nothing the Three Armies can plunder, while the oxen and horses have neither fodder nor place to graze. What should we do?"

The T'ai Kung said: "The Three Armies are unprepared; the oxen and horses have nothing to eat; the officers and troops have no supplies. In this situation seek some opportunity to trick the enemy and quickly get away, setting up ambushes to your rear."

King Wu said: "The enemy cannot be deceived. My officers and troops are confused. The enemy has occupied positions cutting across both our front and rear. Our Three Armies are defeated and in flight. What then?"

The T'ai Kung said: "When you are searching for an escape route, gold and jade are essential. You must obtain intelligence from the enemy's emissaries. In this case cleverness and secrecy are your treasures."

King Wu said: "Suppose the enemy knows I have laid ambushes, so their main army is unwilling to cross the river. The general of their second army then breaks off some units and dispatches them to ford the river. My Three Armies are sorely afraid. What should I do?"

The T'ai Kung said: "In this situation divide your troops into assault formations, and have them improve their positions. Wait until all the enemy's troops have emerged, then spring your concealed troops, rapidly striking their rear. Have your strong crossbowmen on both sides shoot into their left and right flanks. Divide your chariots and cavalry into the Crow and Cloud Formation, arraying them against their front and rear. Then your Three Armies should vehemently press the attack. When the enemy sees us engaged in battle their main force will certainly ford the river and come up. Then spring the ambushing forces, urgently striking their rear. The chariots and cavalry should assault the left and right. Even though the enemy is numerous they can be driven off.

"In general, the most important thing in employing your troops is that when the enemy approaches to engage in battle you must deploy your assault formations, and have them improve their positions. Thereafter divide your chariots and cavalry into the Crow and Cloud Formation. This is the unorthodox in employing your troops. What is referred to as the Crow and Cloud Formation is like the crows dispersing and the clouds forming together. Their changes and transformations are endless."

"Excellent," said King Wu.

49 少 眾

The Few and the Many

KING WU ASKED THE T'AI KUNG: "If I want to attack a large number with only a few, or attack the strong with the weak, what should I do?"

The T'ai Kung said: "If you want to attack a large number with only a few, you must do it at sunset, setting an ambush in tall grass, pressing them on a narrow road. To attack the strong with the weak you must obtain the support of a great state, and the assistance of neighboring states."

King Wu asked: "We do not have any terrain with tall grass, and moreover there are not any narrow roads. The enemy has already arrived; we cannot wait until sunset. I do not have the support of any great state, nor, furthermore, the assistance of neighboring states. What then?"

The T'ai Kung said: "You should set out specious arrays and false enticements to dazzle and confuse their general, to redirect his path so that he will be forced to pass tall grass. Make his route long so you can arrange your engagement for sunset. When his advance units have not yet finished crossing the water, or his rear units have not yet reached the encampment, spring our concealed troops, vehemently striking his right and left flanks, while your chariots and cavalry stir chaos among his forward and rear units. Even if the enemy is numerous, they will certainly flee.

"To serve the ruler of a great state, to gain the submission of the officers of neighboring states, make their gifts generous, and speak extremely deferentially. In this fashion you will obtain the support of a great state, and the assistance of neighboring states."

"Excellent," said King Wu.

50 分 險

Divided Valleys

KING WU ASKED THE T'AI KUNG: "Suppose we have led the army deep into the territory of the feudal lords where we encounter the enemy in the midst of a steep valley. I have mountains on our left, water on the right. The enemy has mountains on the right, water on the left. They divide the valley with us in a standoff. If we choose to defend our position I want to be solid, and victorious if we want to fight. How should we proceed?"

The T'ai Kung said: "If you occupy the left side of a mountain, you must urgently prepare against an attack from the right side. If you occupy the right side of a mountain, then you should urgently prepare against an attack from the left. If the valley has a large river but you do not have boats and oars, you should use the Heavenly Huang to cross the Three Armies over. Those that have crossed should widen the road considerably in order to improve your fighting position. Use the Martial Assault chariots at the front and rear; deploy your strong crossbowmen into ranks; and solidify all your lines and formations. Employ the Martial Assault chariots to block off all the intersecting roads and entrances to the valley. Set your flags out on high ground. This posture is referred to as an Army Citadel.

"In general, the method for valley warfare is for the Martial Assault chariots to be in the forefront, and the Large Covered chariots to act as a protective force. Your skilled soldiers and strong crossbowmen should cover the left and right flanks. Three thousand men will comprise one detachment, which must be deployed in the assault formation. Improve the positions the soldiers occupy. Then the Army of the Left should advance to the left, the Army of the Right to the right, and the Army of the Center to the front, all attacking and advancing together. Those that have already fought should return to their detachment's original positions, the units fighting and resting in succession until you have won."

"Excellent," said King Wu.

VI 犬韜

CANINE
SECRET
TEACHING

51 分合

Dispersing and Assembling

KING WU ASKED THE T'AI KUNG: "If the king, leading the army, has dispersed the Three Armies to several locations and wants to have them reassemble at a specific time for battle, how should he constrain them with oaths, rewards, and punishments so that he can achieve it?"

The T'ai Kung said: "In general, the Way to employ the military, the masses of the Three Armies, must be to have the changes of dividing and reuniting. The commanding general should first set the place and day for battle, then issue full directives and particulars to the generals and commanders setting the time, indicating whether to attack cities or besiege towns, and where each should assemble. He should clearly instruct them about the day for battle, and even the quarter of hour by the water clock. The commanding general should then establish his encampment, array his battle lines, put up a gnomon and erect the official gate, clear the road and wait. When all the generals and commanders have arrived, compare their arrival with the designated time. Those who arrive before the appointed time should be rewarded. Those who arrive afterward should be executed. In this way both the near and distant will race to assemble, and the Three Armies will arrive together, uniting their strength to engage in the battle."

52 武鋒

Military Vanguard

KING WU ASKED THE T'AI KUNG: "In general, when employing the army it is essential to have military chariots, courageous cavalry, a first assault wave, a hand-picked vanguard, and then a perceived opportunity to strike the enemy. In which situations can we strike?"

The T'ai Kung said: "Anyone who wants to launch a strike should carefully scrutinize and investigate fourteen changes in the enemy. Attack when any of these changes become visible, for the enemy will certainly be defeated."

King Wu asked: "May I hear about these fourteen changes?"

The T'ai Kung said: "When the enemy has begun to assemble they can be attacked.

"When the men and horses have not yet been fed they can be attacked.

"When the seasonal or weather conditions are not advantageous to them they can be attacked.

"When they have not secured good terrain they can be attacked.

"When they are fleeing they can be attacked.

"When they are not vigilant they can be attacked.

"When they are tired and exhausted they can be attacked.

"When the general is absent from the officers and troops they can be attacked.

"When they are traversing long roads they can be attacked.

"When they are fording rivers they can be attacked.

"When the troops have not had any leisure time they can be attacked.

"When they encounter the difficulty of precipitous ravines or are on narrow roads they can be attacked.

"When their battle array is in disorder they can be attacked.

"When they are afraid they can be attacked."

53 練士

Selecting Warriors

KING WU ASKED THE T'AI KUNG: "What is the Way to select warriors?"

The T'ai Kung replied: "Within the army there will be men with great courage and strength who are willing to die and even take pleasure in suffering wounds. They should be assembled into a company and called 'Warriors Who Risk the Naked Blade.'

"Those who have fierce *ch'i*, who are robust and courageous, strong and explosive, should be assembled into a company and called 'Warriors Who Penetrate the Lines.'

"Those who are extraordinary in appearance, who bear long swords and advance with measured tread in good order, should be assembled into a company and called 'Courageous, Elite Warriors.'

"Those who can jump well, straighten iron hooks, are powerful, have great strength, and can scatter and smash the gongs and drums, destroy the flags and pennants, should be assembled into a company and called 'Warriors of Courage and Strength.'

"Those who can scale heights and cover great distances, who are light of foot and excel at running, should be assembled into a company and called 'Warriors of the Invading Army.'

"Those who, while serving the ruler, lost their authority and want to again display their merit should be assembled into a company and called 'Warriors Who Fight to the Death.'

"Those who are relatives of slain generals, the sons or brothers of generals who want to avenge their deaths, should be assembled into a company and called 'Warriors Who are Angry unto Death.'

"Those who are lowly, poor, and angry, who want to satisfy their desires, should be assembled into a company and called 'Warriors Committed to Death.'

"Adopted sons and slaves who want to cover up their pasts and achieve fame should be assembled into a company and called the 'Incited Dispirited.'

"Those who have been imprisoned and then spared corporeal punishment, who want to escape from their shame, should be assembled into a company and called 'Warriors Fortunate to be Used.'

"Those who combine skill and technique, who can bear heavy burdens for long distances, should be assembled into a company and called 'Warriors Awaiting Orders.'

"These are the army's selected warriors. You cannot neglect their examination."

54 教戰

Teaching Combat

KING WU ASKED THE T'AI KUNG: "When we assemble the masses of the Three Armies and want to have the officers and men assimilate and become practiced in the Way for teaching combat, how should we proceed?"

The T'ai Kung said: "For leading the Three Armies you must have the constraints of the gongs and drums by which to order and assemble the officers and masses. The generals should clearly instruct the commanders and officers, explaining the orders three times, thereby teaching them the use of weapons, mobilization, and stopping, all to be in accord with the method for changing the flags and signal pennants.

"Thus when teaching the commanders and officers, one man who has completed his study of combat instructions will extend them to ten men. Ten men who have completed their study of combat instructions will extend them to a hundred men. A hundred men who have completed their study of combat instructions will extend them to a thousand men. A thousand men who have completed their study of combat instructions will extend them to ten thousand men. Ten thousand men who have completed their study of combat instructions will extend them to the masses of the Three Armies.

"When the methods of large scale warfare are successfully taught, they will be extended to the masses of millions. In this fashion you will be able to realize a Great Army, and establish your awesomeness throughout the realm."

"Excellent," said King Wu.

55 均兵

Equivalent Forces

KING WU ASKED THE T'AI KUNG: "When chariots and infantry engage in battle, one chariot is equivalent to how many infantrymen? How many infantrymen are equivalent to one chariot? When cavalry and infantry engage in battle, one cavalryman is equivalent to how many infantrymen? How many infantrymen are equivalent to one cavalryman? When chariots and cavalry engage in battle, one chariot is equivalent to how many cavalrymen? How many cavalrymen are equivalent to one chariot?"

The T'ai Kung said: "Chariots are the feathers and wings of the army, the means to penetrate solid formations, to press strong enemies, and to cut off their flight. Cavalry are the army's fleet observers, the means to pursue a defeated army, to sever supply lines, to strike roving forces.

"Thus when chariots and cavalry are not engaged in battle with the enemy, one cavalryman is not able to equal one foot soldier. However, after the masses of the Three Armies have been arrayed in opposition to the enemy, when fighting on easy terrain the rule is that one chariot is equivalent to eighty infantrymen, and eighty infantrymen are equivalent to one chariot. One cavalryman is equivalent to eight infantrymen; eight infantrymen are equivalent to one cavalryman. One chariot is equivalent to ten cavalrymen; ten cavalrymen are equivalent to one chariot.

"The rule for fighting on difficult terrain is that one chariot is equivalent to forty infantrymen, and forty infantrymen are equivalent to one chariot. One cavalryman is equivalent to four infantrymen; four infantrymen are equivalent to one cavalryman. One chariot is equivalent to six cavalrymen; six cavalrymen are equivalent to one chariot.

"Now chariots and cavalry are the army's martial weapons. Ten chariots can defeat a thousand men; a hundred chariots can defeat ten thousand men. Ten cavalrymen can drive off a hundred men, and a hundred cavalrymen can run off a thousand men. These are the approximate numbers."

King Wu asked: "What are the numbers for chariot and cavalry officers and their formations?"

The T'ai Kung said: "For the chariots—a leader for five chariots; a captain for fifteen; a commander for fifty; and a general for a hundred.

"For battle on easy terrain, five chariots comprise one line. The lines are forty paces apart, the chariots from left to right ten paces apart, with detachments sixty paces apart. On difficult terrain the chariots must follow the roads, with ten comprising a company, and twenty a regiment. Front-to-rear spacing should be twenty paces, left-to-right six paces, with detachments thirty-six paces apart. For five chariots there is one leader. If they venture off the road more than a kilometer in any direction, they should return to the original road.

"As for the number of officers in the cavalry—a leader for five men; a captain for ten; a commander for a hundred; a general for two hundred.

"The rule for fighting on easy terrain: five cavalrymen will form one line, and front to back their lines should be separated by twenty paces; left to right four paces; with fifty paces between detachments.

"On difficult terrain the rule is: front to back, ten paces; left to right, two paces. Between detachments, twenty-five paces. Thirty cavalrymen comprise a company; sixty form a regiment. For ten cavalry-men there is a captain. In action they should not range more than a hundred paces, after which they should circle back and return to their original positions."

"Excellent," said King Wu.

56 武車士

Martial Chariot Warriors

KING WU ASKED THE T'AI KUNG: "How does one select warriors for the chariots?"

The T'ai Kung said: "The rule for selecting warriors for the chariots is to pick men under forty years of age, five feet seven inches or taller, whose running ability is such that they can pursue a galloping horse, race up to it, mount it, and ride it forward and back, left and right, up and down, all around. They should be able to quickly furl up the flags and pennants, and have the strength to fully draw an eight-picul cross-bow. They should practice shooting front and back, left and right, until thoroughly skilled. They are termed 'Martial Chariot Warriors.' You cannot but be generous to them."

57 武騎士

Martial Cavalry Warriors

KING WU ASKED THE T'AI KUNG: "How do you select warriors for the cavalry?"

The T'ai Kung said: "The rule for selecting cavalry warriors is to take those under forty, who are at least five feet seven inches tall, strong and quick, who surpass the average. Men who, while racing a horse, can fully draw a bow and shoot. Men who can gallop forward and back, left and right, and all around, both advancing and withdrawing. Men who can jump over moats and ditches, ascend hills and mounds, gallop through narrow confines, cross large marshes, and race into a strong enemy, causing chaos among their masses. They are called 'Martial Cavalry Warriors.' You cannot but be generous to them."

58 戰車

Battle Chariots

KING WU ASKED THE T'AI KUNG: "What about battle chariots?"

The T'ai Kung responded: "The infantry values knowing changes and movement; the chariots value knowing the terrain's configuration; the cavalry values knowing the side roads and the unorthodox Tao. Thus these three armies bear the same name, but their employment differs.

"In general, in chariot battles there are ten types of terrain upon which death is likely, and eight upon which victory can be achieved."

King Wu asked: "What are the ten fatal terrains like?"

The T'ai Kung replied: "If, after advancing, there is no way to withdraw, this is fatal terrain for chariots.

"Passing beyond narrow defiles to pursue the enemy some distance, this is terrain that will exhaust the chariots.

"When the land in front makes advancing easy, while that to the rear is treacherous, this is terrain that will cause hardship for the chariots.

"Penetrating into narrow and obstructed areas from which escape will be difficult: this is terrain on which the chariots may be cut off.

"If the land is collapsing, sinking, and marshy, with black mud sticking to everything, this is terrain that will labor the chariots.

"To the left is precipitous, while to the right is easy, with high mounds and sharp hills. This is terrain contrary to the use of chariots.

"Luxuriant grass runs through the fields, and there are deep watery channels throughout. This is terrain that thwarts the use of chariots.

"When the chariots are few in number, the land easy, and one is not confronted by enemy infantry, this is terrain on which the chariots may be defeated.

"To the rear are water-filled ravines and ditches, to the left deep water, and to the right steep hills. This is terrain upon which chariots are destroyed.

"It has been raining day and night for more than ten days without stopping. The roads have collapsed, so that it is not possible to advance or to escape to the rear. This is terrain that will sink the chariots.

"These ten are deadly terrain for chariots. Thus, they are the means by which the stupid general will be captured and the wise general will be able to escape."

King Wu asked: "What about the eight conditions of terrain that result in victory?"

The T'ai Kung replied: "When the enemy's ranks, front and rear, are not yet settled, strike into them.

"When their flags and pennants are in chaos, their men and horses frequently shifting about, then strike into them.

"When some of their officers and troops advance, while others retreat; when some move to the left, others to the right, then strike into them.

"When their battle array is not yet solid, while their officers and troops are looking around at each other, then strike into them.

"When in advancing they appear full of doubt, and in withdrawing they are fearful, strike into them.

"When the enemy's Three Armies are suddenly frightened, all of them rising up in great confusion, strike into them.

"When you're fighting on easy terrain and twilight has come without being able to disengage from the battle, then strike into them.

"When, after traveling far, at dusk they are encamping and their Three Armies are terrified, strike into them.

"These eight constitute conditions in which the chariots will be victorious.

"If the general is clear about these ten injurious conditions and eight victorious possibilities, then even if the enemy surrounds him on all sides, attacking with a thousand chariots and ten thousand cavalry, he will be able to gallop to the front and race to the sides, and in ten thousand battles invariably be victorious."

"Excellent," said King Wu.

59 戰騎

Cavalry in Battle

KING WU ASKED THE T'AI KUNG: "How should we employ the cavalry in battle?"

The T'ai Kung said: "For the cavalry there are ten situations that can produce victory, and nine that will result in defeat."

King Wu asked: "What are the ten situations that can produce victory?"

The T'ai Kung replied: "When the enemy first arrives and their lines and deployment are not yet settled, the front and rear not yet united, then strike into their forward cavalry, attack the left and right flanks. The enemy will certainly flee.

"When the enemy's lines and deployment are well-ordered and solid, while their officers and troops want to fight, our cavalry should outflank them, but not go far off. Some should race away, some race forward. Their speed should be like the wind, their explosiveness like thunder, so that the daylight becomes as murky as dusk. Change our flags and pennants several times, also change our uniforms. Then their army can be conquered.

"When the enemy's lines and deployment are not solid, while their officers and troops will not fight, press upon them both front and rear, make sudden thrusts on their left and right. Outflank and strike them, and the enemy will certainly be afraid.

"When, at sunset, the enemy wants to return to camp, and their Three Armies are terrified, if we can outflank them on both sides, urgently strike their rear, pressing the entrance to their fortifications, and not allow them to go in, the enemy will certainly be defeated.

"When the enemy, although lacking the advantages of ravines and defiles for securing their defenses, has penetrated deeply and ranged widely into distant territory, if we sever their supply lines, they will certainly be hungry.

"When the land is level and easy, and we see enemy cavalry approaching from all four sides, if we have our chariots and cavalry strike into them, they will certainly become disordered.

"When the enemy runs off in flight, their officers and troops scattered and in chaos, if some of our cavalry outflank them on both sides, while others obstruct them to the front and rear, their general can be captured.

"When at dusk the enemy is turning back while his soldiers are extremely numerous, his lines and deployment will certainly become disordered. We should have our cavalry form platoons of ten, and regiments of a hundred; group the chariots into squads of five and companies of ten; and set out a great many flags and pennants, intermixed with strong crossbowmen. Some should strike their two flanks, others cut off the front and rear, and then the enemy's general can be taken prisoner. These are the ten situations in which the cavalry can be victorious."

King Wu asked: "What about the nine situations that produce defeat?"

The T'ai Kung said: "Whenever the cavalry penetrates the ranks of the enemy but does not destroy their formation, so that the enemy feigns flight only to turn their chariots and cavalry about to strike our rear—this is a situation in which the cavalry will be defeated.

"When we pursue a fleeing enemy onto confined ground, ranging far into their territory without stopping until they ambush both our flanks and sever our rear—this is a situation in which the cavalry will be encircled.

"When we go forward but there is no road back, we enter but there is no way out, this is referred to as 'penetrating a Heavenly Well,' 'being buried in an Earthly Cave.' This is fatal terrain for the cavalry.

"When the way by which we enter is constricted, but the way out is distant; their weak forces can attack our strong ones; and their few can attack our many—this is terrain on which the cavalry will be exterminated.

"When there are great mountain torrents, deep valleys, tall luxuriant grass, forests and trees—these are conditions that will exhaust the cavalry.

"When there is water on the left and right, while ahead are large hills, and to the rear high mountains, and the Three Armies are fighting between the bodies of water, while the enemy occupies both the interior and exterior ground—this is terrain that means great difficulty for the cavalry.

"When the enemy has cut off our supply lines, and if we advance we will not have any route by which to return—this is troublesome terrain for the cavalry.

"When we are sinking into marshy ground, while advancing and retreating must both be through quagmires—this is worrisome terrain for the cavalry.

"When on the left there are deep water sluices, and on the right there are gullies and hillocks, but below the heights the ground appears level, good terrain for advancing, retreating, and enticing an enemy—this terrain is a pitfall for the cavalry.

"These nine comprise fatal terrain for cavalry, the means by which the enlightened general will keep the enemy far off and escape but the ignorant general will be entrapped and defeated."

60 戰 步

The Infantry in Battle

KING WU ASKED THE T'AI KUNG: "What about when infantry engage in battle with chariots and cavalry?"

The T'ai Kung said: "When infantry engage in battle with chariots and cavalry they must rely on hills and mounds, ravines and defiles. The long weapons and strong crossbows should occupy the fore, the short weapons and weak crossbows should occupy the rear, firing and resting in turn. Even if large numbers of the enemy's chariots and cavalry should arrive, they must maintain a solid formation and fight intensely while skilled soldiers and strong crossbowmen prepare against attacks from the rear."

King Wu said: "Suppose there are not any hills or mounds, ravines or defiles. The enemy arrives, and they are both numerous and martial. Their chariots and cavalry outflank us on both sides, and they are making sudden thrusts against our front and rear positions. Our Three Armies are terrified and fleeing in chaotic defeat. What should we do?"

The T'ai Kung said: "Order our officers and troops to set up the *chevaux-de-frise* and wooden caltrops, arraying the oxen and horses by units of five in their midst, and have them establish a four sided martial assault formation. When you see the enemy's chariots and cavalry are about to advance, our men should evenly spread out the caltrops and dig ditches around the rear, making them five feet deep and wide. It is called the 'Fate of Dragon Grass.'

"Our men should take hold of the *chevaux-de-frise* and advance on foot. The chariots should be arrayed as ramparts and pushed forward and back. Whenever they stop set them up as fortifications. Our skilled soldiers and strong crossbowmen should prepare against the left and right flanks. Afterward, order our Three Armies to fervently fight without respite."

"Excellent," said King Wu.